Handbook of Smart Manufacturing Technologies

Smart manufacturing technologies are highly efficient and reliable and produce high-quality products. They involve autonomous and robotic systems, the Internet of Things (IoT), simulation, system integration, and additive manufacturing. These technologies enable industries to make better decisions and respond quickly to changing market conditions. They are used in various fields, including science, medicine, healthcare, and biotechnology. This handbook covers the transformation of conventional to smart manufacturing technologies and the impact of smart manufacturing on industrial engineering.

Handbook of Smart Manufacturing Technologies: Industrial Engineering in the Digital Age covers various smart technologies related to the Internet of Things, robotics, 3D/4D printing, sustainability, and machine learning. It also includes case studies of conventional, advanced, and smart manufacturing, based on real-world aspects. One of the novel aspects of this book is that it provides surface integrity parameters related to machined parts manufacturing and features new product design and development by adapting control algorithms and mechatronics-assisted manufacturing technologies. Furthermore, it offers information on smart manufacturing technologies and their industrial applications.

This handbook is intended to be a useful resource for industrialists, researchers, academicians, or anyone who deals with manufacturing and industrial engineering.

Innovations in Smart Manufacturing for Long-Term Development and Growth

Series Editors: Atul Babbar, Gursel Alici, Yu Dong, Ankit Sharma

Fabrication Techniques and Machining Methods of Advanced Composite Materials
Vikas Dhawan, Atul Babbar, Inderdeep Singh and Jonathan M. Weaver

Advances in Pre- and Post-Additive Manufacturing Processes
Innovations and Applications
Naveen Mani Tripathi and Ankit Sharma

Handbook of 3D Printing in Pharmaceutics
Innovations and Applications
Prakash Katakam, Ranvijay Kumar, Nishant Ranjan and Atul Babbar

Intelligent Machinery Fault Diagnostics and Prognostics
The Future of Smart Manufacturing
Deepam Goyal, Ankit Sharma and Mohamad Abou Houran

Nanomaterials for Smart Manufacturing
Fundamental Principles, Theory, Case Studies, and Advanced Applications
Dinesh Deshwal, Anil Kumar Narwal, Vivek Srivastava, Atul Babbar and Jonathan M. Weaver

Sensor Technologies and Wireless Communications in Industry 5.0
A Solution for Smart Manufacturing
Edited by Ramkumar K.R., Bhupendra Singh, Lei Pan and Ankit Sharma

Handbook of Smart Manufacturing Technologies
Industrial Engineering in the Digital Age
Edited by Mahendra U. Gaikwad, Ankit Sharma, Muthuramalingam Thangaraj and Sachin Salunkhe

For more information about this series, please visit: www.routledge.com/Innovations-in-Smart-Manufacturing-for-Long-Term-Development-and-Growth/book-series/CRCISMLDG

Handbook of Smart Manufacturing Technologies

Industrial Engineering in the Digital Age

Edited by
Mahendra U. Gaikwad, Ankit Sharma,
Muthuramalingam Thangaraj, and
Sachin Salunkhe

CRC Press
Taylor & Francis Group
Boca Raton London New York

CRC Press is an imprint of the
Taylor & Francis Group, an informa business

Designed cover image: Shutterstock

First edition published 2026
by CRC Press
2385 NW Executive Center Drive, Suite 320, Boca Raton FL 33431

and by CRC Press
4 Park Square, Milton Park, Abingdon, Oxon, OX14 4RN

CRC Press is an imprint of Taylor & Francis Group, LLC

ISBN: 978-1-032-81979-2 (hbk)
ISBN: 978-1-032-82037-8 (pbk)
ISBN: 978-1-003-50264-7 (ebk)

DOI: 10.1201/9781003502647

Typeset in Times
by Apex CoVantage, LLC

Dr. Mahendra U. Gaikwad dedicates this book to his affectionate parents, Late Shri Uttam Ambadas Gaikwad (Ex–Assistant Labor Officer and Public Relations Officer, Solapur Municipal Corporation, Maharashtra, India) and Smt. Subhadra Uttam Gaikwad (Ex–High School Teacher, Bharti Vidyapeeth Madhyamik Vidyalaya and Junior College, Solapur, Maharashtra, India), whose values and sacrifices laid the foundation for all his endeavors. To his loving wife, Mayuri Gaikwad, whose unwavering encouragement and trust have been his greatest sources of strength and inspiration. To his beloved children, Mr. Siddharth Gaikwad and Ms. Rajnandini (Vaishnavi) Gaikwad, and to his bright and sincere nephews, Mr. Utkarsh Gaikwad and Mr. Prasenjit Gaikwad, whose love, support, and encouragement made this book project a successful reality.

Dr. Ankit Sharma dedicates this book to Dr. Harshita Bhargave (wife) and Mishka Sharma (daughter).

In the end, we would also like to dedicate this book to all the scientific, global collaboration and research communities.

Contents

Preface......xi
About the Editors......xiii
Contributors......xv
Acknowledgments......xix

SECTION I Transformation of Conventional to Smart Manufacturing Technologies in the Digital Age

Chapter 1 Advances in Hybrid Metal Additive Manufacturing Technology and the Challenges for Manufacturers, Software, and Professionals......3

Marco Aurélio Feriotti, Alexandre Formigoni, Rosinei Batista Ribeiro, Eliacy Cavalcanti Lélis, Fabrício José Piacente, Caio Fernando da Silva, and Ramón Quiza Sardiñas

Chapter 2 Influence of Various Process Parameters on Bead Geometry, Mechanical Properties, and Microstructural Characteristics of Produced WAAMed Parts in Wire Arc Additive Manufacturing (WAAM) of Metallic Alloys: A Comprehensive Review......21

Swarup Mandal, Rajesh Porwal, and Manidipto Mukherjee

Chapter 3 Teaching Learning–Based Optimization for Optimizing Process Parameter in Wire Electrical Discharge Machining (WEDM)......40

Shalaka Kulkarni, Omkar K. Kulkarni, Samidha A. Jawade, Swanand Pachpore, and Ganesh Kakandikar

Chapter 4 Peering into the Smart Manufacturing—Material Forming and Joining: Casting and Welding to Unbind the Potential of Traditional Manufacturing......59

Sreejith S. Nair, Mahendra U. Gaikwad, and Kirubanidhi Jebabalan

Chapter 5 Computational and Experimental Analysis of Erosion Wear in a Pipeline with Butterfly Valves 85

Laerson Nhantumbo, Gurmeet Singh, and Harjot Singh Gill

Chapter 6 Enhancing Concrete Performance with Sugarcane Bagasse Ash: Impact on Workability, Strength, and Durability 99

Mohammad Tabish, Mohd Moonis Zaheer, Abdul Baqi, and Salim Barbhuiya

SECTION II Transformation of Advanced Materials Processing Toward Industrial Engineering

Chapter 7 Technological Prospecting for the Sugarcane Juice Extraction Stage: A Patentometric Study on the Diffuser and Crushing Mill Process 113

Fabrício José Piacente, Silva Vanessa de Cillos, Biaggi Denys Eduardo, Alexandre Formigoni, Rosinei Batista Ribeiro, Eliacy Cavalcanti Lélis, and Marco Aurélio Feriotti

Chapter 8 Comparative Analysis of Recycled HDPE Tiles with Graphene against Solar Panels: A Study from the Perspective of Frugal Innovation and Technoscience as an Instrument for Technological and Social Development 126

Sergio Assis Machado, Rosinei Batista Ribeiro, Adison da Silva Mello, Francisco Del Moral Hernandez, Caio Fernando da Silva, and Ramón Quiza Sardiñas

Chapter 9 Feasibility Study and Design of Heat Sink of Passive Cooling System 142

Jaswinder Singh Mehta

Chapter 10 Effects of the Addition of Plastic Waste on the Strength Properties of Concrete and Bituminous Asphalt 154

Ranjodh Singh and Ravpreet Kaur

Chapter 11 Using Value Stream Mapping (VSM) as a Lean Production Tool: Streamlining the Production Process in a Lubricant and Degreaser Company 168

Erik Leonel Luciano, Rosinei Batista Ribeiro, Eliane Antônio Simões, Alexandre Formigoni, and Marcos Antonio Bessa

Chapter 12 Study of Design Requirements for the Design of Bakelite Materials for Smart Manufacturing Applications 179

Caio Fernando da Silva, Rosinei Batista Ribeiro Marco Aurélio Feriotti, Alexandre Formigoni, Marcelino Rivas Santana, Eliane Antônio Simões, and Karolyne Borges de Oliveira

Index 191

Preface

This book emerges as a comprehensive volume for manufacturing professionals, researchers, and academicians seeking knowledge in the domain of Smart Manufacturing Technologies and Industrial Engineering in the Digital Age. As industries transition towards intelligent, adaptive, and efficient manufacturing processes, the convergence of advanced manufacturing, artificial intelligence, IoT, signal processing, and computer vision plays a pivotal role in the digital age.

The first chapter highlights the 'Advances in Hybrid Metal Additive Manufacturing Technology and the Challenges for Manufacturing Professionals'. Further, the second chapter focuses on the 'Influence of Process Parameters on Wire Arc Additive Manufacturing (WAAM) Process for Metallic Alloys'. Chapter 3 is oriented to optimizing process parameters in wire electrical discharge machining. This chapter deals with the influence of various input parameters, which include pulse-off time, pulse-on time, wire tension, wire feed, lower flush, and upper flush, to obtain the optimum result of the surface roughness.

Chapter 4, 'Peering into the Smart Manufacturing-Material Forming and Joining: Casting and Welding to Unbind the Potential of Traditional Manufacturing', highlights the challenges of adopting smart technologies in traditional manufacturing processes, like metal forming, casting, and welding. It also explores how these processes have undergone transformative changes, powered by smart technologies and digitalization in the era of Industry 4.0. Chapter 5 explained the computational and experimental analysis of erosion wear in pipeline applications. Further, Chapter 6 outlines the impact on Workability, Strength, and Durability on the Concrete Performance with Sugarcane Bagasse Ash. In Chapter 7, a patentometric study on the diffuser and crushing mill process has been elaborated while considering the sugarcane juice extraction stage. In the investigation study of Chapter 8, Comparative Analysis of Recycled HDPE Tiles with Graphene Against Solar Panels has been discussed. Chapter 9 highlights the Design of the heat sink of the passive cooling system. Chapter 10 described the effects of the Addition of Plastic Waste on the Strength Properties of Concrete and Bituminous Asphalt. Along with it, Chapter 11 elaborates on the application of Value Stream Mapping as a Lean Production Tool to further streamline the Production process in the Lubricants. The final chapter unfolds a different facet of the Bakelite material for smart manufacturing applications.

This book also underscores the importance of cross-disciplinary integration. Smart manufacturing requires collaboration across diverse fields, and our exploration reflects this interconnectedness. Whether you are an engineer, researcher, or educator, the insights provided here cater to your needs, ensuring relevance and applicability in your respective domains. We hope this book serves as a beacon of knowledge, guiding you towards innovative solutions and transformative advancements in your professional endeavors.

About the Editors

Dr. Mahendra U. Gaikwad is Assistant Professor in the Department of Production Engineering, Veermata Jijabai Technological Institute (VJTI), Mumbai (an Autonomous Institute of the Government of Maharashtra), India. He obtained his B.E. in Mechanical Engineering and M.E. in Mechanical Design Engineering from the Walchand Institute of Technology, Solapur, Maharashtra, India, and Ph.D. in Mechanical Engineering from the Sathyabama Institute of Science and Technology, Chennai, Tamil Nadu, India. His research interests include advanced manufacturing processes, material characterization, surface integrity, and machining of smart material alloys and optimization. In addition to having over 12 years of teaching experience at the UG and PG levels, Dr. Gaikwad has published more than 25 articles and 3 book chapters; edited 2 books, including *Advance Materials: Production, Characterization and Multidisciplinary Applications* (CRC Press/Taylor & Francis); and has been a reviewer for multiple international journals and international conferences. Dr. Gaikwad has undertaken industry-based research projects for UG and PG students and has guided more than 25 UG project groups and over 8 PG students in his teaching career. In 2022, he received the Best Performer in Education (Teaching & Research Excellence) Award from the Savitribai Phule Excellence Award, Pune, Maharashtra, India. He is a member of the International Association of Engineers (IAENG) and a lifetime member of the Quality Circle Forum of India.

Dr. Ankit Sharma is Associate Director (Research) at the Center for Research, Impact and Outcome at Chitkara University, Punjab, India. He received his doctorate in Mechanical Engineering from the Thapar Institute of Engineering and Technology, India. He has more than ten years of experience in academics, research, consulting, training, and industry. Dr. Sharma has authored numerous national and international publications, has filed/published over 30 international patents, and is the editor of the CRC Press/Taylor & Francis book series Innovations in Smart Manufacturing for Long-Term Development and Growth. He has been an editor of multiple books related to advanced manufacturing processes published by CRC Press/Taylor & Francis and is the managing and special issue editor of several international journals with databases of SCIE, ESCI, and Scopus-indexed. He has also delivered several invited seminars and keynote talks on international (USA, China, and India) platforms and was awarded with best research paper awards. He is also on the reviewer board of several international journals.

Dr. Muthuramalingam Thangaraj is Professor and Head in the Department of Mechatronics Engineering, SRM Institute of Science and Technology, Kattankulathur Campus, India. He has vast experience both in Academic and Industry. His primary area of research interests are unconventional machining processes and Mechatronics with good hands on experience. He has published many of his research papers in the area of EDM, ECM, AWJM, LBM processes, micro machining, 3D Printing,

Flexible Printed sensors, optimisation approaches and Manufacturing automation and control. He has been listed as Top 2% scientists in worldwide by Stanford University, USA since 2020. He has published around 200 research articles and 8 patents so far. HE has presented his research works in various International conferences held around the world. Among which 3 patents are granted so far. He has been involved with various national accreditation activities, Academic council members, research forum members, research funding evaluation committee member for the benefits of students. He serves for many reputed journals in various positions such as editors and reviewers. He represented Engineering Education in terms of visiting Professor, MoU signs, Keynote speech, Session chair and paper presentation in many International Conferences and competitions held in Abroad countries such as USA, UK, Germany, Belgium, China, UAE, Netherland, Turkey, Poland, Greece, Austria, Czech republic, Slovakia, Hungary, Malaysia, Singapore, Thailand, Sri Lanka, Saudi Arabia and Vietnam under various schemes. He is actively involved for establishing academic and research collaboration with reputed universities. He has involved in many research finding as Technical Investigator around the world.

Dr. Salunkhe Sachin is Professor in the Mechanical Engineering Department at Gazi University, Ankara, Turkey. He received his bachelor's in mechanical engineering from Dr. Babasaheb Ambedkar Marathwada University, Aurangabad, India, his Master of Technology in CAD/CAM, and his Ph.D. from S.V. National Institute of Technology, Surat, India. He has more than 120 research papers in reputed journals and conferences and 9 book chapters to his credit. Also, Dr. Sachin is an editor of various reputable journals from various publishers. His main research interests are in press tool design, metal additive manufacturing, AI applications in manufacturing, expert systems for die design, sheet metal forming, wear analysis, fatigue analysis, and high strain rate.

Contributors

Abdul Baqi
Department of Civil Engineering, Zakir Husain College of Engineering and Technology, Aligarh Muslim University
Aligarh, India

Salim Barbhuiya
Department of Engineering and Construction, School of Architecture, Computing and Engineering
University of East London
London, UK

Marcos Antonio Bessa
ANASER Industrial
São Paulo, Lorena, Brazil

Eliacy Cavalcanti Lélis
Unidade de Pós-Graduação, Extensão e Pesquisa (UPEP) do Centro Estadual de Educação Tecnológica Paula Souza—CEETEPS
Brazil
Centro Universitário Teresa D'Ávila—UNIFATEA
Brazil

Caio Fernando da Silva
Federal University of Itajubá, Graduate Program in Development, Technology and Society, Itajubá Minas Gerais, Brazil

Adison da Silva Mello
Federal University of Itajubá, Graduate Program in Development, Technology and Society
Itajubá Minas Gerais, Brazil

Silva Vanessa de Cillos
FATEC—Piracicaba, Centro Estadual de Educação Tecnológica Paula Souza—CEETEPS
Brazil

Karolyne Borges de Oliveira
Universidade Federal de Itajubá (UNIFEI)
Itajubá, Minas Gerais, Brazil
Centro Universitário Teresa D'Ávilla (UNIFATEA)
Lorena, São Paulo, Brazil

Biaggi Denys Eduardo
ETEC—Americana, Centro Estadual de Educação Tecnológica Paula Souza—CEETEPS
Brazil

Marco Aurélio Feriotti
Unidade de Pós-Graduação, Extensão e Pesquisa (UPEP) do Centro Estadual de Educação Tecnológica Paula Souza—CEETEPS
Brazil

Alexandre Formigoni
Unidade de Pós-Graduação, Extensão e Pesquisa (UPEP) do Centro Estadual de Educação Tecnológica Paula Souza—CEETEPS
Brazil

Mahendra U. Gaikwad
Department of Production Engineering, Veermata Jijabai Technological Institute (VJTI)
Mumbai, Maharashtra, India

Harjot Singh Gill
Department of Mechanical Engineering, Chandigarh University
Punjab, India

Francisco Del Moral Hernandez
Paula Souza State Center for Technological Education, Postgraduate, Extension and Research Unit—CEETEPS—UPEP
São Paulo, São Paulo, Brazil

Samidha A. Jawade
Mechanical Engineering, MIT—World Peace University Pune
India

Kirubanidhi Jebabalan
Department of Industrial Engineering, Division of Material Science, University of Trento
Povo 2 Campus, Italy

Fabrício José Piacente
Unidade de Pós-Graduação, Extensão e Pesquisa (UPEP) do Centro Estadual de Educação Tecnológica Paula Souza—CEETEPS
Brazil

Ganesh Kakandikar
Mechanical Engineering, MIT—World Peace University
Pune, India

Ravpreet Kaur
Shree Guru Gobind Singh Tricentenary University
Gurugram, India
I. K. Gujral Punjab Technical University,
Jalandhar, Punjab, India

Omkar K. Kulkarni
Mechanical Engineering, MIT—World Peace University
Pune, India

Shalaka Kulkarni
Mechanical Engineering, MIT—World Peace University
Pune, India

Erik Leonel Luciano
Postgraduate, Extension and Research Unit (UPEP)—Paula Souza State Center for Technological Education—CEETEPS
São Paulo, Brazil

Sergio Assis Machado
Federal University of Itajubá, Graduate Program in Development, Technology and Society
Itajubá, Minas Gerais, Brazil

Swarup Mandal
Regional Science City (A unit of NCSM)
Lucknow, U.P., India

Jaswinder Singh Mehta
UIET, Panjab University
Chandigarh, India

Manidipto Mukherjee
CSIR Central Mechanical Engineering Research Institute (CMERI)
Durgapur, W.B., India

Sreejith S. Nair
Department of Mechanical Engineering, Dr. Mahalingam College of Engineering and Technology, Pollachi
Coimbatore, Tamil Nadu, India

Laerson Nhantumbo
Department of Mechanical Engineering, Chandigarh University
Punjab, India

Swanand Pachpore
Awanti Polymoulds Pvt. Ltd
Pune, India

Rajesh Porwal
Shri Ramswaroop Memorial University Lucknow
Barabanki, U.P., India

Rosinei Batista Ribeiro
Paula Souza State Center for Technological Education, Postgraduate, Extension and Research Unit—CEETEPS—UPEP, São Paulo
São Paulo, Brazil

Marcelino Rivas Santana
Centro de Estudios de Fabricación Avanzada y Sostenible de la Universidad de Matanzas (CEFAS-UM) Matanzas, Cuba

Ramón Quiza Sardiñas
Universidad de Matanzas—Centro de Estudios de Fabricación Avanzada y Sostenible—CEFAS
University of Matanzas—Center of Advanced and Sustainable Manufacturing Studies—UMCC—CEFAS, Cuba

Eliane Antonio Simões
Centro Estadual de Educação Tecnológica Paula Souza—CEETEPS—São Paulo
São Paulo, Brazil

Gurmeet Singh
Department of Mechanical Engineering, Chandigarh University
Punjab, India

Ranjodh Singh
DAV University
Punjab, India

Mohammad Tabish
Department of Civil Engineering, Zakir Husain College of Engineering and Technology, Aligarh Muslim University
Aligarh, India

Mohd Moonis Zaheer
Department of Civil Engineering, Zakir Husain College of Engineering and Technology, Aligarh Muslim University
Aligarh, India

Acknowledgments

We express our sincere thanks to all contributors, all those friends and colleagues who have activated our norms to take up a study like this.

We extend our profound gratitude to our institutions, whose unwavering support and dynamic research ecosystems have been instrumental in shaping this work. Dr. Mahendra U. Gaikwad would like to express his sincere appreciation to Dr. Sachin D. Kore, Director, and Dr. R. N. Awale, Deputy Director Veermata Jijabai Technological Institute (VJTI), Mumbai, Maharashtra, India, for their consistent motivation, guidance, and support throughout the successful journey of this book. Last but not least, he wishes to convey his heartfelt gratitude to Dr. Dadarao N. Raut, Professor & Head of the Production Engineering Department, VJTI, and Dr. Madhukar R. Nagare, Associate Professor,Production Enginnering Department, VJTI for their unwavering encouragement and invaluable support in his research endeavors.

We owe a deep debt of thanks to our families, whose enduring patience, unwavering encouragement, and the nurturing environments they cultivated at home have been the true backbone of this endeavor. Their silent strength and sacrifice have been the invisible yet indispensable force driving us forward.

We also express our sincere appreciation to CRC Press, Taylor and Francis for its professional guidance and support throughout the publication process, ensuring the global reach and impact of this work.

Mahendra U. Gaikwad
Ankit Sharma
Muthuramalingam Thangaraj
Sachin Salunkhe

Section I

Transformation of Conventional to Smart Manufacturing Technologies in the Digital Age

1 Advances in Hybrid Metal Additive Manufacturing Technology and the Challenges for Manufacturers, Software, and Professionals

Marco Aurélio Feriotti, Alexandre Formigoni, Rosinei Batista Ribeiro, Eliacy Cavalcanti Lélis, Fabrício José Piacente, Caio Fernando da Silva, and Ramón Quiza Sardiñas

1.1 INTRODUCTION

Integrating additive and subtractive manufacturing (SM) in the same machine tool with computer numerical control (CNC) is called hybrid metal additive manufacturing (HMAM). This combination provides the manufacturing of metal parts with better finishing, shorter processing time, and greater efficiency, reducing errors caused when transferring the printed part for finishing on other equipment. These characteristics make HMAM a very interesting technological advancement (Atkinson, 1985).

Given this circumstance and the possibility of producing complex metallic parts with exclusive properties, this technology has aroused interest in the industrial sector, which finds a range of possibilities (Rave, 2011). However, innovation carries relevant challenges related to the emergence of equipment, the insufficiency of options for specific software, and the scarcity of qualified professionals to operationalize the technology, maximizing its potential (Mordor Intelligence, 2022).

This work focuses on HMAM, the combination of additive manufacturing (AM) of metals with the machining process, highlighting the challenges related to equipment manufacturers, the lack of adequate software, and the need for specialized professionals for the effectiveness of its implementation.

It is important to recognize the obstacles associated with implementing HMAM, despite its numerous benefits. The questions at the forefront of this study explored the

DOI: 10.1201/9781003502647-2

following: "What are the main obstacles that impede HMAM advances, particularly in terms of equipment with advanced technology, specific software integrating AM with machining, and qualified professionals?"

In this context, the general objective of this study is to verify the advancement of HMAM, noting the challenges relevant to equipment manufacturers, software options, and the training of specialized professionals.

The specific objectives to achieve this general objective are:

- Check the advances in equipment recently developed for HMAM.
- Investigate the shortage of specialized software and workable solutions to this gap.
- Explore strategies for training professionals to work in this field and retaining talent.

The aerospace, automotive, and medical industries can be transformed if production is upgraded, but HMAM is not necessarily focused on this. There are some problems surrounding this technology that must be overcome if we want to achieve better results. Addressing the challenges and capitalizing on the opportunities presented by HMAM is the ultimate goal of this investigation. Valuable information will be provided to various stakeholders such as manufacturers, educational institutions, and software developers.

This chapter is structured into four sections: Introduction—presentation and contextualization of the research topic, and a summary of the theory used to develop the research; Methodology—presents the choice of the research method; Results and Discussions—contains the analysis of data collected during the research in the databases; Final Considerations—presents the conclusion and limitations of the research carried out and suggests future studies.

1.2 HYBRID METAL ADDITIVE MANUFACTURING

Early advances in HMAM focused on the use of multiple thermal energies and the integration of metal additive manufacturing (MAM) with metal cutting processes, aiming to improve productivity and the quality of the produced parts. These advances led to the commercialization of the first HMAM systems in the mid-2010s (Pragana et al., 2021).

Recently, the integration of MAM with metal forming processes has emerged, with the initial goal of improving the shape of the deposited material layer through localized plastic deformation while simultaneously increasing the stiffness and wear resistance of the manufactured part (Sealy et al., 2018).

This hybrid approach, which combines MAM with forming, has developed rapidly in recent years. MAM is divided into seven application processes and is currently used in the manufacture of metal parts (Figure 1.1) (Pragana et al., 2021).

This study focuses on direct energy deposition (DED), which produces metal parts by delivering powder or wire through a nozzle into the part being built, where the material is melted by a concentrated source of thermal energy (Fidan et al., 2024). In this process, raw materials are supplied in powder form from a feeder and melted

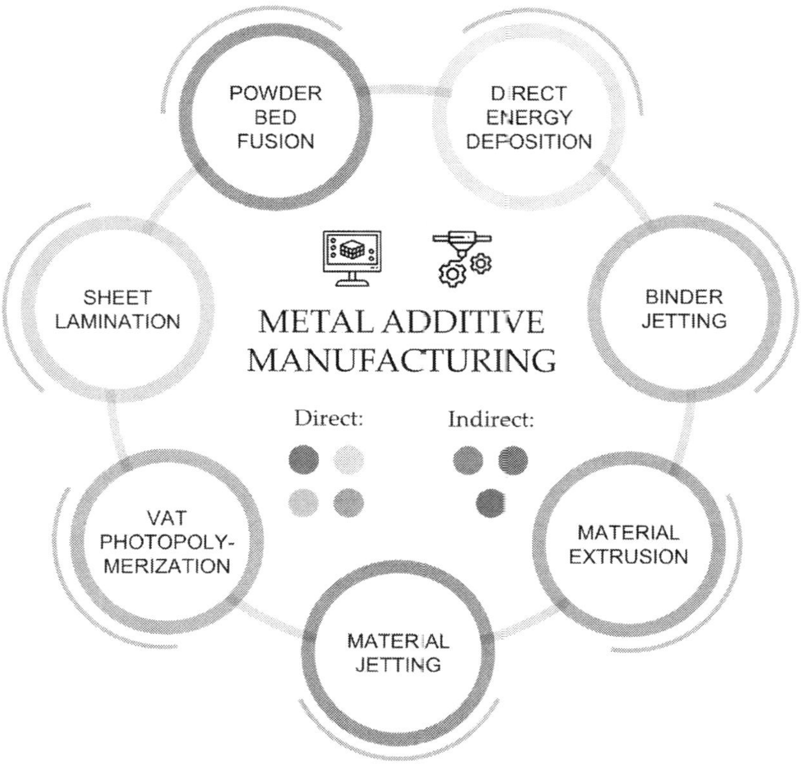

FIGURE 1.1 Application processes of metal additive manufacturing.

(Pragana et al., 2021)

precisely as they are deposited, rather than remaining stationary on the build platform while the part is formed layer by layer, as shown in the schematic representation in Figure 1.2 (Pragana et al., 2021).

The integration of DED processes with CNC machining machines, known as HMAM, has given rise to hybrid manufacturing (HM) systems that combine milling, turning, and MAM capabilities, as shown in the schematic representation in Figure 1.3 (Pragana et al., 2021).

These systems enable the production of complex, off-the-shelf metal parts using a single platform and shorten lead times (Lorenz et al., 2015; Merklein et al., 2016).

HMAM combines AM (3D printing) with conventional machining processes to create complex metal components. This technology provides several advantages, including greater design freedom, the ability to produce parts with specific properties and characteristics, such as wear resistance, and better electrical and thermal conductivity (Posada et al., 2019).

Eliminating the need to transfer the printed part to a separate machine for finishing saves time, reduces errors, and results in greater efficiency. This is one of the main

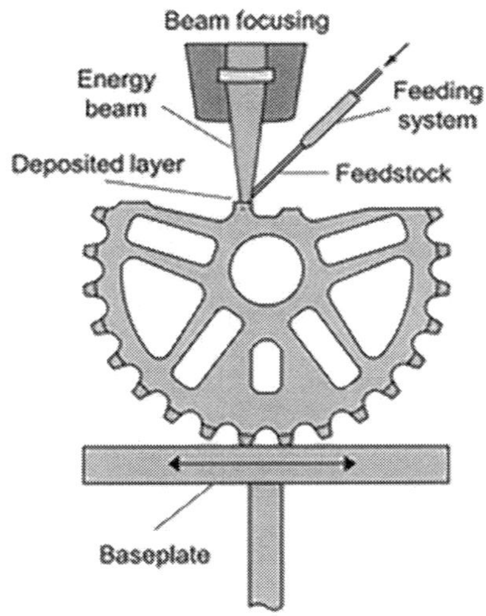

FIGURE 1.2 Schematic representation of DED.

(Adapted from Pragana et al., 2021)

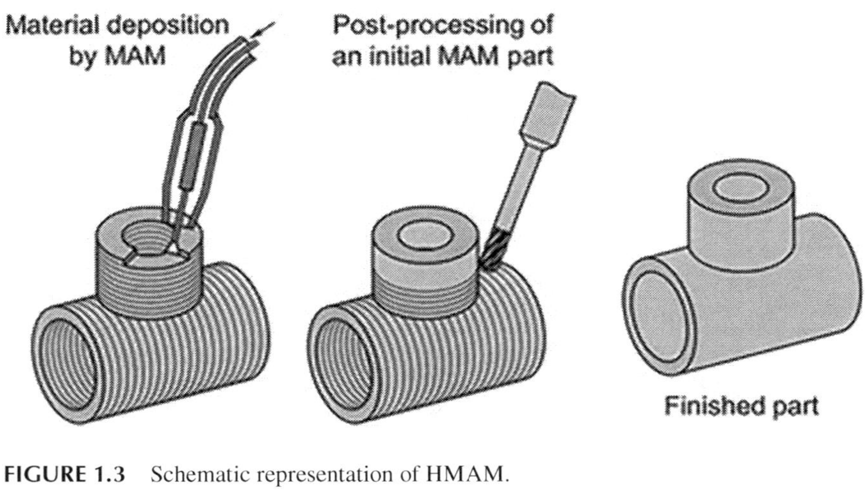

FIGURE 1.3 Schematic representation of HMAM.

(Adapted from Pragana et al., 2021)

advantages of combining additive and conventional processes in a single machine (Vafadar et al., 2021).

At HMAM, it is possible to use multiple materials in the same part. This flexibility in applying materials with different mechanical properties makes it possible to add steel with greater hardness or copper to improve heat transfer (Dilberoglu et al., 2021). The ability to add material only when necessary to repair damaged parts or reduce machining labor can save costs and increase the sustainability of the manufacturing process (Feriotti et al., 2024).

1.3 ADVANCES IN MANUFACTURING EQUIPMENT

Recent advancements in AM have garnered substantial industrial interest (Sharma et al., 2018, 2019a, 2019b, 2020, 2021, 2022a, 2022b).

Initially, AM was predominantly linked to prototype fabrication; however, ongoing technological developments, coupled with an expanding variety of available materials—particularly for metallic components, have broadened its scope of applications. As a result, AM is now increasingly utilized for the production of functional parts as well (Jiménez et al., 2021).

Previous research has found that knowledge in this area is increasing in the number of international publications. This is due to the growing importance of this topic in educational institutions and industries around the world (Feriotti & Formigoni, 2022).

Another conclusion that can be inferred is that there is no specific trend in the adoption of this technology; that is, it can be used in various sectors, as there are a large number of different countries and sectors working on HM and its applications (Feriotti & Formigoni, 2022).

Equipment manufacturers play a key role in advancing HMAM. They invest in research and development to increase the yield, speed, and reliability of their manufacturing systems. Furthermore, new metallic materials compatible with this technology are continually being developed, further expanding its range of applications (Cortina et al., 2018).

Due to the complexity of the MAM process, system manufacturers have developed a set of optimized processing conditions for some existing metal powders. Table 1.1 defines specific machine models and manufacturers for one or more MAM powders (Asnafi, 2021).

AM technologies enable the creation of complex and topologically optimized geometries, including internal cavities that are unattainable with traditional manufacturing processes (Gaikwad et al., 2025). However, achieving strict geometric tolerances and meeting stringent surface integrity requirements are often not feasible with standalone AM technologies (Jiménez et al., 2021)

Consequently, as emphasized by Fidan et al. (2023), AM-produced parts often require extensive post-processing to meet stringent surface and dimensional requirements, as well as to achieve the desired mechanical properties. In this context, the integration of AM with post-processing technologies into single- or multi-configuration processing solutions—commonly referred to as hybrid AM—has emerged as an attractive proposition for the industry, while also garnering significant interest in research and development (Jiménez et al., 2021).

Hybrid manufacturing machines, developed by companies such as Matsuura, DMG Mori, ROMI HYBRID, and Hermle, offer advanced features and solutions for diverse manufacturing needs, making HM a viable option for industries looking to harness the benefits of additive and subtractive processes (Vafadar et al., 2021).

TABLE 1.1
Configuration of Machine Models and Manufacturers

Manufacturer	Model	Al	Co	Cu	Fe	Ni	Ti	W
3D Systems	DMP Factory 500 PrinterModule	By request				Nickel alloys	By request	
	DMP Factory/Flex 350	AlSi7Mg0.6, AlSi10Mg	CoCrF75	—	Maraging steel, 17-4PH, 316L	Ni625, Ni718	Ti Gr1, Ti Gr5, Ti Gr23	—
	DMP Flex 100	—	CoCr	—	17-4PH, 316L	—	—	—
	ProX DMP 300	AlSi12	CoCr	—	Maraging steel, 17-4PH	—	—	—
	ProX DMP 200	AlSi12	CoCr	—	Maraging seel, 17-4PH, 316L	—	—	—
Additive Industries	MetalFAB1	AlSi10Mg, ScalmAlloy©	—	—	Tool steel 1.2709, 316L	IN718	Ti6Al4V	—
Concept Laser	X Line 2000R	AlSi10Mg	Balanced productivity	—	—	Nickel 718	Ti6AL4V, Grade 23	—
	M Line Factory	A205	CoCrMo	—	—	Nickel 718CL	—	—
	M2 Multilaser	AlSi10Mg, AlSi7Mg	CoCrMo	—	Maraging M300, 17-4PH, 316L	Nickel 625, Nickel 718	Ti6AL4V, Grade 23	—
EOS	EOS M 400-4	AlSi10Mg	—	—	MS1, 316L	HX, IN718	Ti64, Ti64ELI	
	EOS M 400	AlSi10Mg	—	—	MS1, 316L	IN718	Ti64	
	EOS M 300-4	AlSi10Mg	—	—	MS1, 316L	IN718		
	EOS M 290	AlSi10Mg	MP1	—	MS1, CX, PH1, 17-4PH, 316L	HX, IN625, IN718	Ti64, Ti64ELI, TiCP Grade 2	
	EOS M 100	—	SP2	—	316L	—	Ti64	W1

SLMSolutions	SLM®NXG XII 600	ALSi10Mg (no limitation)		No limitation		IN718 (Unlimited)	No limitation	
	SLM®800 SLM®500 SLM®280 SLM®125	AlSi10Mg, AlSi7Mg0.6, AlSi9Cu3	CoCr28Mo6, SLM® MediDent	CuSn10, CuNi2SiCr	Maraging 1.2709, 316L (1.4404), 15-5PH (1.4545), 17-4PH (1.4542), Invar 36®	HX, IN625, IN718, IN939	Ti6Al4V ELI (Grade 23), TA15, e Ti (Grade 2)	—
Renishaw	RenAM 500Q/S RenAM 500E RenAM 500M AM 400 AM 250	AlSi10Mg	CoCr	—	Maraging M300, 316L	IN625, IN718	Ti6Al4ELI	—
Trumpf	TruPrint 1000	Yes to all except W+ precious metal alloys + amorphous metals						
	TruPrint 2000	Yes to all except Cu and W+ amorphous metals						
	TruPrint 3000	Yes to all except Co, Cu, and W						
	TruPrint 5000	Yes to all except Co, Cu, and W						

1.4 SPECIFIC SOFTWARE

The absence of computer-aided process planning systems constrains the efficient use of additive-subtractive HM in various industrial applications due to the complexity involved in designing system inputs during the design phase. Therefore, the development of software for additive-subtractive hybrid systems is crucial to encourage further implementation and expand the adoption of AM in the future (Abdulhameed et al., 2019).

The software used in hybrid AM plays a fundamental role in creating the CNC commands that control the machine during the 3D printing process and subsequent machining. This software is responsible for translating the three-dimensional digital model of the part to be manufactured into precise instructions for the CNC machine (Farias & Vilarinho, 2022).

The lack of connectivity, combined with the limitation of MAM processes in producing finished parts in a single step, underscores the importance of integrating these processes with upstream and downstream operations. This integration has the potential to significantly reduce overall production time, while also enhancing the precision and quality of the produced parts (Panagiotis et al., 2018).

Simulating the process in the software is an important step that precedes sending the CNC code to the machine. This allows you to check the machining or deposition virtually and identify potential problems or collisions that may occur during actual operation (Posada et al., 2019).

The software used to generate programs for DED in HM is that which allows the integration of additive and subtractive processes in a single environment, simulating and verifying operations and generating G-code for hybrid machines (Santoro, 2015).

The lack of specialized software is one of the main obstacles faced by HMAM. The integration of 3D printing processes with machining requires robust software that can improve production, ensuring quality and efficiency. Currently, few comprehensive solutions are available, making this a critical area of development (Thompson et al., 2016).

1.5 SHORTAGE OF TRAINED PROFESSIONALS

When the focus is on increasing productivity or innovation processes, learning can serve as a basis for organizations and universities to absorb more knowledge (Feriotti & Formigoni, 2022).

For a professional to work with hybrid AM of metals, several technical competencies and specific skills are necessary, due to the complexity and interdisciplinary nature of this area. Understanding the properties of metallic materials is fundamental to selecting the appropriate materials for each application and understanding how they behave during manufacturing and machining (Beekman, 2022).

HMAM requires professionals to be skilled in computer-aided manufacturing (CAM) software to create efficient tool paths and generate CNC code, in addition to knowing how to integrate these processes with 3D printing, understanding the principles of AM, including different 3D printing techniques and parameter configuration, such as laser power, material flow, deposition speed, and path strategy. Be able to identify and solve technical problems and challenges related to HMAM (Cortina et al., 2018).

Successful HMAM implementation requires an advanced team. Professionals with knowledge in part design for AM, machine programming, and machining processes are essential. However, training and attracting talent to this area are still significant challenges (Barbosa et al., 2019).

1.6 METHODOLOGY

To achieve the objectives of this study, reviews of scientific literature related to HMAM, AM equipment and software manufacturers, and training of professionals were conducted.

According to Leão (2019), to become familiar with the phenomenon or obtain a new understanding of it, scientific publications were investigated, and as a strategy the search for scientific articles, in English, in the Web of Science and Scopus databases, published between 2017 and 2023, was conducted. The choice of these databases is based on their breadth and widespread use.

The investigation of virtual environments as a field of study has been widely adopted, especially in qualitative research. Observation and document analysis represent essential methodological strategies in this context. The dynamics of online social interactions offer a rich and relevant research object for observation, while the vast amount of information and data available in these environments creates a well-established documentary collection for analysis and research (Salgueiro et al., 2020).

In this context, specific websites in the area were also used as sources of information, considering selection criteria, such as the relevance of the websites to the research topic, the credibility of the sources, and the updating of the information. The choice of academic, institutional websites, professional organizations, and specialized forums stands out; outdated and unreliable information was excluded. The searches were conducted using keywords related to the research objective and Google Chrome as the engine.

The evaluation of the data approach found on the current state of equipment development, the software available for this application, and the demand for qualified professionals to work in this area allowed a broad understanding of the current challenges and trends in this field.

1.7 RESULTS AND DISCUSSIONS

The collected data were interpreted, and the results obtained were contextualized about the general objective of this research. In this section, the relevant conclusions derived from the investigation into HMAM advances were explored, as well as the meaningful evaluation of these results. This favored understanding existing literature, practical applications, and the development of knowledge in this constantly evolving field.

1.7.1 Advances in Equipment

In the context of HM, the ability to predict outcomes during both the design and production phases is critically important. Ongoing efforts are focused on achieving greater operational efficiency and minimizing material waste through modeling programs and simulation techniques. The integration of partial intelligent systems is essential for addressing future challenges associated with "Industry 4.0" (Mehmet et al., 2022).

The aerospace and automotive sectors have been adopting the use of HMAM technology, thanks to notable advances by equipment manufacturers in developing more accessible and efficient systems. The application of real-time monitoring technologies, multi-material printing systems, and greater machining precision are the factors that drive the adoption of the technology.

Due to several dominant businesses in the region and the significant use of HM machines in various industries, North America has contributed significantly to the revenue of hybrid AM machines (Mordor Intelligence, 2022).

According to Heghe and Sirois (2023), the continued growth of the aerospace and automobile industries in the United States is accelerating the industry's development. Growing awareness about HM machines, which offer production efficiency, is driving global demand in the industry.

The competitive environment of the hybrid AM machines market presents a moderate concentration, due to the presence of large suppliers that offer the machines and equipment used in HM. These suppliers hold the largest market share, as depicted in Figure 1.4 (Heghe & Sirois, 2023).

Companies that provide HM machines adopt a variety of strategies, including new product development, innovations, mergers and acquisitions, partnerships, and collaborations (Heghe & Sirois, 2023). These actions aim to maintain their position in the highly competitive market and add HM machine offerings on a global scale, with

FIGURE 1.4 Hybrid additive manufacturing machine market—concentration global by region, 2016–2023.

(https://www.mordorintelligence.com/industry-reports/hybrid-additive-manufacturing-machine-market)

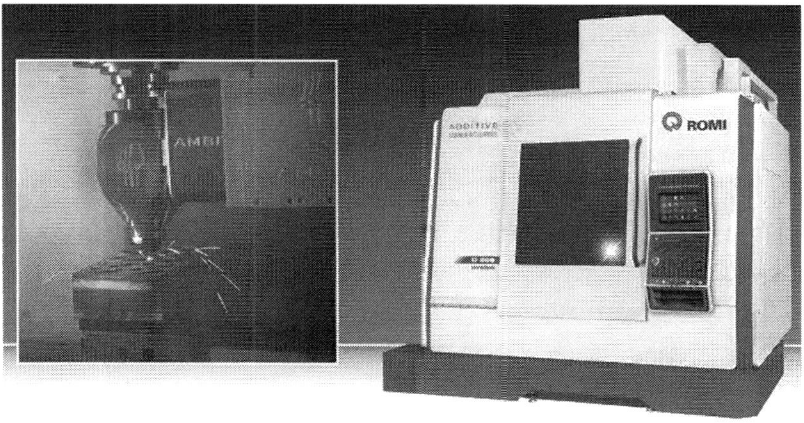

FIGURE 1.5 ROMI D800 HYBRID machining center.
(https://www.romi.com/produtos/manufatura-hibrida-romi/)

examples of partnerships such as ROMI (Brazilian company) and HYBRID (UK Company).

In the ROMI D 800 HYBRID Machining Center (Figure 1.5), it is feasible to conduct the material deposition and machining steps in a single fixture, which provides greater precision to the final part and improves the efficiency of the production process, reducing the handling time of the piece. Regarding its application versatility, the ROMI Hybrid Line can serve a wide range of sectors, including molds and dies, oil and gas, aeronautics, healthcare, and automotive industry, and is frequently adopted in the research centers (Raissa, 2021).

The global hybrid AM machines market, estimated at US$2.4 billion in the year 2022, is expected to reach US$9.2 billion by 2030, growing at a compound annual growth rate of 20.2% during the period from 2023 to 2030. The US market is estimated at US$740.6 million, while China, Japan, Canada, and Germany are expected to grow during 2023–2030, as shown in Figure 1.6 (Heghe & Sirois, 2023).

By the year 2030, the AM and materials market is projected to expand to a value of US$167.50 billion. It is widely recognized by companies from different sectors that AM is a crucial technology to be incorporated into their factories (Heghe & Sirois, 2023).

Numerous advantages and inefficiencies associated with these manufacturing techniques are manifested in factors such as design, fabrication, material properties, and working conditions. Consequently, the development of a production technology that integrates the benefits of both conventional and additive techniques is of substantial significance (Mehmet et al., 2022).

Awareness and collaborative efforts among experts from various disciplines, such as materials science, mechanics, chemistry, economics, and management, are essential for fostering a sustainable future in terms of the product lifecycle. This involves evaluating the supply chain footprint of AM and employing methodologies necessary to compare all stages of the process, from material extraction to the final

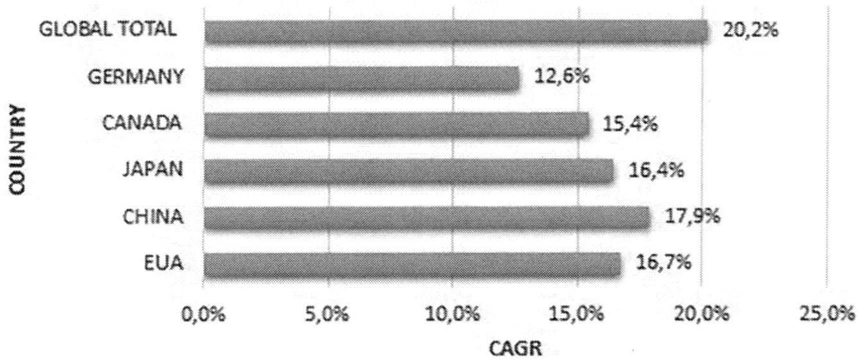

FIGURE 1.6 Estimated global growth of the hybrid additive manufacturing machines market until 2030.

(Adapted from Heghe & Sirois, 2023)

product (Feriotti et al., 2022). Integrating these perspectives is crucial for developing production technologies that combine the benefits of both conventional and additive techniques, thereby maximizing efficiency and minimizing environmental impact.

1.7.2 Specific Software

Although there have been advances in creating specific software for HMAM, there is still a significant gap. Integrating 3D printing and machining workflows requires more robust, interconnected solutions. Software companies are starting to address this problem, but there is still much work to be done. In this research, two software programs were found available for application at HMAM:

ESPRIT CAM: This software allows you to create DED additive cycles with customizable parameters, simulate and verify additive construction, integrate subtractive processes, and generate optimized G-code for hybrid machines. ESPRIT CAM supports several hybrid machine manufacturers, such as DMG Mori, Mazak, Okuma, and Sodick.

Siemens NX: This software allows you to model and simulate additive and subtractive processes in an integrated environment, define deposition and machining strategies, validate part quality, and generate G-code for hybrid machines. Siemens NX supports several hybrid machine manufacturers, such as DMG Mori, Mazak, Trumpf, Optomec, and ROMI HYBRYD.

ESPRIT CAM and Siemens NX are both CAM programming software that enable the user to automate, speed up, and optimize the production of complex parts. However, there are some differences between them, as described in Table 1.2.

TABLE 1.2
Differences between ESPRIT CAM and Siemens NX Software

Resources	ESPRIT CAM	Siemens NX
Compatibility	Compatible with the Meltio hybrid manufacturing engine, which allows additive and subtractive manufacturing methods to be combined in a single machine.	Does not have this integration but offers advanced additive manufacturing capabilities with NX hybrid additive manufacturing
Functionality	Focused only on CAM but can import data from CAD and other formats.	Integrated design, engineering, and manufacturing that includes CAD, CAE, and CAM capabilities
Licensing Options	The pricing for ESPRIT CAM is not available online, but a personalized quote can be requested.	Offers various licensing options, including perpetual or subscription-based licenses and basic, machine, or advanced packages
Source	https://www.3dcadportal.com/dp-technology-soporta-la-manufactura-aditiva-ded.html	https://blogs.sw.siemens.com/nx-manufacturing/hybrid-additive-manufacturing-with-nx-cam-overview/

The choice between these software options will depend on the specific needs of each user and the priorities regarding integration, functionalities, scope, and licensing model.

The hardware of hybrid AM systems is currently undergoing continuous advancement. For instance, multi-axis AM/SM machine tools enable process planners to utilize non-uniform slice thicknesses and adjust surface normal directions through in-process machining of the base slice (Ugur et al., 2019).

Although it is anticipated that the development of accompanying software tools will keep pace with these advancements, there have been limited efforts reported in the computer-aided planning of hybrid process combinations based on AM (Ugur et al., 2019).

Therefore, developing new virtual AM/SM manufacturing systems or integrating AM into existing CAM systems is crucial for realizing supportive tools for hybrid AM. A recent study has distinctly identified the unique challenges and requirements for hybrid CAM systems, despite the current lack of implementations (Ugur et al., 2019).

Given that conventional machine tools and additive platforms operate using compatible machine codes (NC-code/G-code) to produce the target object, hybrid CAM software must also function with similar codes. Indeed, generating machine-specific instructions is a well-established task for SM with cutting-edge technologies. Extending this task to hybrid AM/SM workstations is a straightforward process (Ugur et al., 2019).

1.7.3 Shortage of Trained Professionals

The lack of professionals remains a crucial challenge. Educational institutions and companies need to collaborate to create training programs that prepare engineers and technicians to work with hybrid AM of metals. In addition, talent retention is important, as competition for specialists is intense (Thompson et al., 2016).

Training professional agents for HMAM becomes more challenging since operators will need to acquire comprehensive knowledge to master the various processes involved. The use of computational tools to optimize the hybrid process requires commitment and significant investment in training (Barbosa et al., 2019).

There are a few educational institutions in Brazil offering training for professionals working in the field of HMA:

The National Industrial Learning Service (SENAI) Joinville offers a professional extension course in Additive Manufacturing in Metals, focusing on the technique of selective laser melting. The course develops the skills to carry out analysis, planning, and support for the execution of products by additive manufacturing in metals. (Sesi Senai, 2023)

The Institute for Technological Research (IPT) has a research center in Additive Manufacturing of Metals, which develops projects with different metal alloys in partnership with other institutions such as Poli-USP, ITA, and Fraunhofer. (Raissa, 2021)

The joint action with the São Carlos School of Engineering University of São Paulo (EESC-USP), the São Paulo Research Foundation (FAPESP), and Hybrid Manufacturing aims to train and educate professionals through research (Instituto de Pesquisas Tecnológicas, n.d.).

Although various studies on HAM are being conducted by software developers and machine manufacturers, the same cannot be said for the barrier of limited expertise. In the case of AM, academic curricula often overlook these technologies, and the same applies to lifelong learning programs, leading to a shortage of qualifications in higher education, vocational education, and training, as well as appropriate qualification and certification methodologies for AM (Panagiotis et al., 2023).

AM is driving profound changes across the entire manufacturing product chain; thus, simply integrating an AM machine into an existing manufacturing setup is insufficient for successful application—it is merely the beginning. Successful integration of AM into the future manufacturing landscape necessitates comprehensive training for the next generation of professionals (Panagiotis et al., 2023).

1.8 FINAL CONSIDERATIONS

HMA aims to play an increasingly significant role in the manufacturing industry. Advances in equipment are promising and continue to make this technology more accessible. However, the shortage of suitable software and the lack of trained professionals represent major challenges that need to be overcome to reach its full potential.

Equipment manufacturers, software companies, and educational institutions have a responsibility to collaborate in solving these challenges. As more resources are devoted to research and development in these areas, we expect the adoption of HMA to improve and a revolution in how complex metal parts are designed and manufactured.

The future of manufacturing is inseparable from the success of HMA, and further investment is required to realize this potential.

HMA technology can benefit a variety of industries but tends to have a greater impact on industries that require complex metal parts with customized properties. Industries most interested in this technology include:

Aerospace: HMA enables the manufacturing of complex and lightweight aerospace parts, improving fuel consumption efficiency in aircraft. Examples include turbine blades, supports, and structural components.

Automotive: Hybrid manufacturing in the automotive sector involves rapid prototyping, customized and durable parts, and using multi-materials to improve vehicle performance, reducing fuel consumption and pollutant emissions.

Medicine: In the medical and health sectors, the use of HMA allows the production of personalized implants and specific surgical instruments, reducing the risk of rejection and complications, and bringing substantial benefits to patients.

Energy: Hybrid manufacturing creates custom parts for critical equipment in the energy industry, including oil and gas. These designs improve component efficiency and durability, reducing maintenance costs and downtime.

Defense: HMA is applied in the military industry to create lightweight and highly resistant components. This shortens delivery times on critical projects and allows rapid prototyping of special components for validation.

Tools and molds: In this sector, HMA is used to recover molds and dies, as well as to create inserts with optimized cooling ducts. The use of multi-materials promises to improve wear resistance and extend the life of the mold, reducing the need for maintenance.

Electronics: Hybrid manufacturing in the electronics industry makes it possible to produce heat sinks, connectors, and complex components with excellent thermal conductivity and high electrical properties.

Dentistry and orthopedics: Through HMA, dental implants, crowns, and personalized orthopedic prostheses can be manufactured according to the patient's anatomy, benefiting the dental and orthopedic areas.

Jewelry: Hybrid manufacturing is a leading choice among jewelry designers looking to create sophisticated, personalized jewelry due to the varied metal options and ability to produce pieces with complex geometries that are difficult to manufacture conventionally.

Research and development: The development of HMA has been taking place within research institutions and laboratories in some sectors that, through experiments and the development of prototypes, are seeking to advance knowledge of materials and understanding of manufacturing processes.

It was observed that technology is constantly evolving, as studies and research progress, its applications expand to new sectors, opening new possibilities for its implementation in industries.

Based on a survey of HMA's technological advances, a careful analysis highlighted its impact on the industry. This chapter presents a gap in the technological challenges for its implementation, among which the few specific software options and the lack of specialized professionals stand out.

Due to the complexity and interdisciplinary nature of the area, educational institutions need to train professionals capable of working in HMA. This requires the development of specific skills in AM, metallurgy, design, equipment, software, safety, and regulations. Furthermore, it is essential to have knowledge of materials and an understanding of the properties of metals used in AM.

The CAM software described earlier needs to be able to integrate the metal deposition process (AM) with machining, allowing programming and simulations combining both technologies. This integration is critical to creating parts that are constructed using AM technology and then machined to meet final tolerances and specifications.

This chapter brings significant considerations about HMA and its challenges for researchers, industry professionals, and political decision-makers, aiming to stimulate and direct future research in this constantly evolving area, as well as broader debates.

REFERENCES

Abdulhameed, O., Al-Ahmari, A., Ameen, W., & Mian, S. H. (2019). Additive Manufacturing: Challenges, Trends, and Applications. Advances in Mechanical Engineering, 11(2). https://https://doi.org/10.1177/1687814018822880

Asnafi, N. (2021). Application of Laser-Based Powder Bed Fusion for Direct Metal Tooling. Metals, 11(3), 458. https://doi.org/10.3390/met11030458

Atkinson, A. O. (1985). Design for Manufacturability: Computer-Integrated Design and Manufacturing for Product Development (SAE Technical Paper 851587). SAE International. https://doi.org/10.4271/851587

Barbosa, F. A., Velázquez, D. R. T., Helleno, A. L., & Junior, M. V. (2019, novembro 6). Vantagens e Desafios da Manufatura Hibrida—Integrando Manufatura Aditiva e Subtrativa. ENEGEP 2019—Encontro Nacional de Engenharia de Produção, SANTOS, SP, BRASIL. https://doi.org/10.14488/ENEGEP2019_TN_WPG_290_1638_38258

Beekman, J. (2022, julho 18). Opportunities and Benefits of Hybrid Manufacturing. IoT Marketing. https://iotmktg.com/opportunities-and-benefits-hybrid-manufacturing/

Cortina, M., Arrizubieta, J. I., Ruiz, J. E., Ukar, E., & Lamikiz, A. (2018). Latest Developments in Industrial Hybrid Machine Tools That Combine Additive and Subtractive Operations. Materials, 11(12), Artigo 12. https://doi.org/10.3390/ma11122583

Curso Manufatura Aditiva em Metais—Extensão Profissional—SESI SENAI Joinville SC. (2023). http://cursos.sesisenai.org.br/detalhes/manufatura-aditiva-em-metais/10776

Dilberoglu, U. M., Gharehpapagh, B., Yaman, U., & Dolen, M. (2021). Current Trends and Research Opportunities in Hybrid Additive Manufacturing. The International Journal of Advanced Manufacturing Technology, 113(3), 623–648. https://doi.org/10.1007/s00170-021-06688-1

Farias, R. M., & Vilarinho, L. O. (2022). Simulações Computacionais dos Processos de Manufatura Aditiva de Metais: Um Review Introdutório. Soldagem & Inspeção, 27, e2703. https://doi.org/10.1590/0104-9224/SI27.03

Feriotti, M. A., & Formigoni, A. (2022). Análise Bibliométrica Sobre a Tendência da Aplicação da Manufatura Híbrida. South American Development Society Journal, 8(24), Artigo 24. https://doi.org/10.24325/issn.2446-5763.v8i24p64-80

Feriotti, M. A., Formigoni, A., Silva, C. F. da, Ribeiro, R. B., & Simões, E. A. (2024). Applicability of Hybrid Metal Manufacturing (HMM), Characteristics, Steel Types, and Mechanical Properties: Systematic Literature Review. IOSR Journal of Humanities and Social Science, 29, 7–21. https://doi.org/10.9790/0837-2908100721

Ferrotti, M. A., Pohlmann, M. N., Formigoni, A., & Oliveira, R. de F. (2022). Manufatura Aditiva e Sustentabilidade: Estudo Exploratório para Avaliar Potenciais Impactos Ambientais. South American Development Society Journal, 8(24), 148. https://doi.org/10.24325/issn.2446-5763.v8i24p148-166

Fidan, I., Alshaikh Ali, M., Naikwadi, V., Gudavasov, S., Mahmudov, M., Mohammadizadeh, M., . . . & Sharma, A. (2024). Nano-Level Additive Manufacturing: Condensed Review of Processes, Materials, and Industrial Applications. Technologies, 12(7), 117.

Fidan, I., Huseynov, O., Ali, M. A., Alkunte, S., Rajeshirke, M., Gupta, A., . . . & Sharma, A. (2023). Recent Inventions in Additive Manufacturing: Holistic review. Inventions, 8(4), 103.

Gaikwad, M. U., Gaikwad, P. U., Ambhore, N., Sharma, A., & Bhosale, S. S. (2025). Powder Bed Additive Manufacturing Using Machine Learning Algorithms for Multidisciplinary Applications: A Review and Outlook. Recent Patents on Mechanical Engineering, 18(1), 12–25. http://dx.doi.org/10.2174/0122127976289573240319102303

Heghe, J. V., & Sirois, C. (2023). Global Hybrid Additive Manufacturing Machine Market Size Study & Forecast, by End-User Industry (Aerospace & Defense, Energy & Power, Electronics, Medical, Automotive, and Other End-User Industries), and Regional Analysis, 2022–2029. https://www.giiresearch.com/report/bzc1236126-global-hybrid-additive-manufacturing-machine.html

Instituto de Pesquisas Tecnológicas. (n.d.). MAM/IPT – Manufatura Aditiva de Metais. https://mam.ipt.br

Jiménez, A., Bidare, P., Hassanin, H., et al. (2021). Powder-Based Laser Hybrid Additive Manufacturing of Metals: A Review. The International Journal of Advanced Manufacturing Technology, 114, 63–96. https://doi.org/10.1007/s00170-021-06855-4

Leão, L. M. (2019). Metodologia do estudo e pesquisa: Facilitando a vida dos estudantes, professores e pesquisadores. Editora Vozes.

Lorenz, K. A., Jones, J. B., Wimpenny, D. I., & Jackson, M. R. (2015). A review of Hybrid Manufacturing. Solid Freeform Fabrication Conference Proceedings, 53, 96–108.

Mehmet, E. K., Saad, W., Garcia-Collado, A., Munish, K. G., & Grzegorz, M. K. (2022). A Technical Overview of Metallic Parts in Hybrid Additive Manufacturing Industry. Journal of Materials Research and Technology, 18, 384–395. https://doi.org/10.1016/j.jmrt.2022.02.085

Merklein, M., Junker, D., Schaub, A., & Neubauer, F. (2016). Hybrid Additive Manufacturing Technologies—an Analysis Regarding Potentials and Applications. Physics Procedia, 83, 549–559. https://doi.org/10.1016/j.phpro.2016.08.057

Mordor Intelligence. (2022). Mercado de Máquinas de Manufatura Aditiva Híbrida | 2022–27 | Participação da indústria, tamanho, crescimento—Mordor Intelligence. https://www.mordorintelligence.com/industry-reports/hybrid-additive-manufacturing-machine-market

Panagiotis, S., Panagis, F., Alexios, P., & Harry, B. (2018). Addressing the Challenges for the Industrial Application of Additive Manufacturing: Towards a Hybrid Solution. International Journal of Lightweight Materials and Manufacture, 1, 157–168. https://doi.org/10.1016/j.ijlmm.2018.07.002

Panagiotis, S., Panagis, F., John, S., & Harry, B. (2023). Increasing the Industrial Uptake of Additive Manufacturing Processes: A Training Framework. Advances in Industrial and Manufacturing Engineering, 6. https://doi.org/10.1016/j.aime.2022.100110

Posada, A., Rodriguez, E., & Alvares, A. (2019). A Functional Application Activities Model for Hybrid Additive-Subtractive Manufacturing. Anais do 10º Congresso Brasileiro de Engenharia de Manufatura, São Carlos, SP, Brasil.

Pragana, J. P. M., Sampaio, R. F. V., Bragança, I. M. F., Silva, C. M. A., & Martins, P. A. F. (2021). Hybrid Metal Additive Manufacturing: A State-of-the-Art Review. Advances in Industrial and Manufacturing Engineering, 2, 100032. https://doi.org/10.1016/j.aime.2021.100032

Raissa. (2021, maio 3). Manufatura Aditiva Metálica para criar réplica da Taça Jules Rimet. ROMI S.A. https://www.romi.com/manufatura-aditiva-metalica-para-criar-replica-da-taca-jules-rimet/

Rave, J. I. P. (2011). El avión de la muda: Herramienta de apoyo a la enseñanza-aprendizaje práctico de la manufactura esbelta. Revista Facultad de Ingeniería Universidad de Antioquia, (58), 173–182.

Salgueiro, J. B., Silva, V. L. M. da, & Silva, C. S. P. (2020). Orientações sobre Ética em Pesquisa em Ambientes Virtuais. Comitê de Ética em Pesquisa. Escola Nacional de Saúde Pública Sergio Arouca (ENSP/Fiocruz).

Santoro, D. (2015, novembro 2). Hybrid Additive Manufacturing with NX CAM Overview—NX Manufacturing. https://blogs.sw.siemens.com/nx-manufacturing/hybrid-additive-manufacturing-with-nx-cam-overview/

Sealy, M. P., Madireddy, G., Williams, R. E., Rao, P., & Toursangsaraki, M. (2018). Hybrid Processes in Additive Manufacturing. Journal of Manufacturing Science and Engineering, 140(6). https://doi.org/10.1115/1.4038644

Sharma, A., Babbar, A., Tian, Y., Pathri, B. P., Gupta, M., & Singh, R. (2022b). Machining of Ceramic Materials: A State of the Art Review. International Journal on Interactive Design and Manufacturing, 17(6), 2891–2911.

Sharma, A., Jain, V., & Gupta, D. (2018). Characterization of Chipping and Tool Wear during Drilling of Float Glass Using Rotary Ultrasonic Machining. Measurement, 128, 254–263.

Sharma, A., Jain, V., & Gupta, D. (2019a). Comparative Analysis of Chipping Mechanics of Float Glass during Rotary Ultrasonic Drilling and Conventional Drilling: For Multi-Shaped Tools. Machining Science and Technology, 23, 547–568.

Sharma, A., Jain, V., & Gupta, D. (2019b). Multi-Shaped Tool Wear Study during Rotary Ultrasonic Drilling and Conventional Drilling for Amorphous Solid. Proceedings of the Institution of Mechanical Engineers, Part E: Journal of Process Mechanical Engineering, 233, 551–560.

Sharma, A., Jain, V., & Gupta, D. (2021). Effect of Pre and Post Tempering on Hole Quality of Float Glass Specimen: For Rotary Ultrasonic and Conventional Drilling. Silicon, 13, 2029–2039.

Sharma, A., Jain, V., & Gupta, D. (2022a). Mathematical Approach on Chipping Volume Estimation Generated during Rotary Ultrasonic Drilling for Float Glass. Proceedings of the National Academy of Sciences, India Section A: Physical Sciences, 92, 285–291.

Sharma, A., Jain, V., Gupta, D., & Babbar, A. (2020). A Review Study on Miniaturization. In Advanced Manufacturing and Processing Technology (First edition, pp. 111–131). Boca Raton, FL: CRC Press.

Thompson, M. K., Moroni, G., Vaneker, T., Fadel, G., Campbell, R. I., Gibson, I., Bernard, A., Schulz, J., Graf, P., Ahuja, B., & Martina, F. (2016). Design for Additive Manufacturing: Trends, Opportunities, Considerations, and Constraints. CIRP Annals: Manufacturing Technology, 65(2), 737–760. https://doi.org/10.1016/j.cirp.2016.05.004

Ugur, M. D., Vahid, H., Ulas, Y., & Melik, D. (2019). Simulator of an Additive and Subtractive Type of Hybrid Manufacturing System. Procedia Manufacturing, 38, 792–799. https://doi.org/10.1016/j.promfg.2020.01.110

Vafadar, A., Guzzomi, F., Rassau, A., & Hayward, K. (2021). Advances in Metal Additive Manufacturing: A Review of Common Processes, Industrial Applications, and Current Challenges. Applied Sciences, 11(3), Artigo 3. https://doi.org/10.3390/app11031213

2 Influence of Various Process Parameters on Bead Geometry, Mechanical Properties, and Microstructural Characteristics of Produced WAAMed Parts in Wire Arc Additive Manufacturing (WAAM) of Metallic Alloys

A Comprehensive Review

Swarup Mandal, Rajesh Porwal, and Manidipto Mukherjee

2.1 INTRODUCTION

2.1.1 Additive Manufacturing

The process of additive manufacturing (AM) of metals, also termed '3D metal printing', builds items by incrementally adding material layer by layer. The AM process, which has evolved over the past three decades, is widely used to produce intricately shaped metal objects and is becoming increasingly popular each year (Gaikwad et al., 2025). Currently, the objective of research work on the AM process has shifted toward the ability to construct complicated design metallic parts (Ding et al., 2015). Initially, it produces complicated design metal parts that are challenging or are not feasible

DOI: 10.1201/9781003502647-3

to produce with conventional methods (Tripathi & Sharma, 2024). Additionally, this method significantly reduces wastage of material by maintaining precise geometrical accuracy. Therefore, the benefits of utilizing the AM process are more noticeable when producing smaller quantities, as the unit costs are somewhat lower than those of traditional manufacturing methods (Sharma et al., 2018, 2019a, 2019b, 2020, 2021, 2022, 2023).

2.1.2 WAAM

This part introduces the WAAM process, its evaluation with other powder and wire-based systems, and its various applications for metallic AM. WAAM has become increasingly popular today due to the enhanced accumulation and increased levels of freedom in power sources and how they are controlled. The wire arc fabrication equipment can achieve the desired properties at different sites by adjusting different deposition parameters of AM like current, voltage, wire feed speed (WFS), and travel speed (TS). The WAAM process specifically depends on the gas metal arc welding (GMAW) technique and has the potential to build large objects because of its ability to provide a higher deposition rate. Arc welding is preferred over electron AM and laser AM processes because it is more economical and also faster.

In AM, as laid stress by Fidan et al. (2024), wire and arc components are manufactured by applying a bead of welded metal layer by layer using GMAW welding and using a positioning system like a CNC positioning system. The main benefit of utilizing the WAAM process is its high deposition rate, leading to decreased production and lead times.

Figure 2.1 shows a typical WAAM system consisting of an arc welding machine, a wire feeder, a computer control unit, a robotic arm or a computerized CNC system, a welding torch, a sensor unit, and a substrate. Since WAAM primarily requires a hardware setup, which includes an arc welding machine and a robotic arm, the focus has predominantly been on the hardware aspects, leading to the development of a variety of hardware systems by researchers. During the WAAM process, complex physical phenomena occur, causing significant variations in the manufactured components under different technological conditions. Exploring the detailed physical mechanisms of the WAAM process is essential (Li et al., 2022). Additionally, integrating a sensor system is typically necessary to monitor molding information and enable real-time control during the process. Detailed exploration of the physical mechanisms underlying the WAAM process is essential. Moreover, the incorporation of a sensor system is typically essential during the process to monitor molding information effectively and facilitate real-time control. It becomes essential to carry out an evaluation of the properties exhibited by the additive part. This evaluation is crucial in determining the standard of the end product.

2.1.3 WAAM Setup

The deposition work has been performed by utilizing a unique type of welding setup to fabricate the thin-walled samples, as presented in Figure 2.2. This setup consists of the following devices: a six-axis robotic handling system, welding machine/power source, shielding gas, filler material, and substrate material.

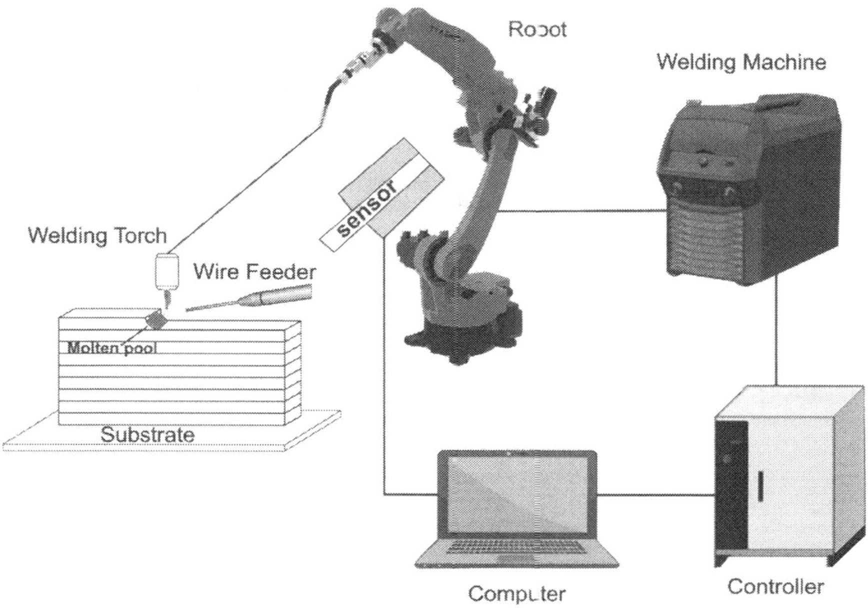

FIGURE 2.1 Representation of the WAAM system.

FIGURE 2.2 Typical WAAM setup at CSIR-CMERI, Durgapur.

2.2 THE GENERAL PROSPECT OF THE WAAM

2.2.1 WAAM Working Principle

The WAAM process is a fusion-based manufacturing technique that utilizes heat generated by an electric arc to melt metal wire feedstock. This molten wire feedstock metal substance is then deposited on a substrate plate layer by layer to create structures. By precisely layering this molten material, WAAM can create complex geometries and build composite structures by cladding different materials in a single process.

2.2.2 Types of WAAM

As per ASTM F42 standards, a committee based on AM technologies was formed in 2009 by ASTM, which provides comprehensive guidelines and classifies the WAAM process into seven categories. Among these, the first four categories are particularly appropriate for use with metals or metallic alloys.

1. Power Bed Fusion
2. Sheet Lamination
3. Binder Jetting
4. Directed Energy Deposition
5. Vat Photo Polymerization
6. Material Jetting
7. Material Extrusion

The WAAM process used in this study can be incorporated into the direct energy deposition (DED) methods, where concentrated thermal energy is applied to melt and fuse materials during deposition.

Depending on the nature of the heat source used, DED techniques can be categorized into three sections: laser, electron beam, and electric arc. Electric arc serves as the heating source in the WAAM process and metal as the deposition material. The WAAM processes can be classified as gas metal arc–directed energy deposition (GMA-DED), gas tungsten arc–directed energy deposition (GTA-DED), and plasma arc–directed energy deposition (PA-DED) processes based on the source of heat energies. Figures 2.3 and 2.4 represent the schematic diagram of the welding-based AM process and the classifications and brief descriptions of each process.

(a) s+GMA-DED or GMAW

The welding process known as GMA-DED is an electric arc generated between a usable wire electrode and the workpiece metal and is widely employed in WAAM processes. This process is also known as GMAW. During this WAAM process, an electric arc forms in the middle of the usable wire electrode's tip and the metal part being manufactured, safeguarded by inert or active shielding gases. Utilizing a continuous spool and nozzle, the wire is fed automatically, melting into droplets that are separated and integrated within the molten pool formed by the electric arc.

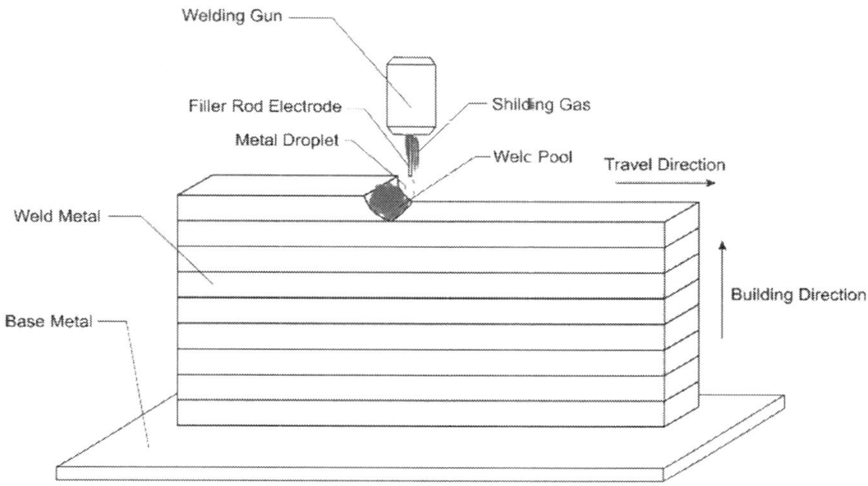

FIGURE 2.3 Welding-based additive manufacturing.

In the GTA-DED and PA-DED processes, a tungsten non-consumable electrode is used to form an electric arc with the metallic component. It is also known as gas tungsten arc welding (GTAW) and plasma arc welding (PAW). Unlike GMA-DED, the wire feed direction in GTA-DED and PA-DED can vary, impacting the quality of the deposit and complicating the planning of the process.

2.2.3 WAAM Process Parameters and Their Influences on Weld Beads

WAAM is recognized as a type of DED process where the metallic materials are deposited layer by layer using an arc heat source to form a 3D structure from a predetermined 3D CAD model. The layer deposition, however, required two major pre-requisite responses, such as layer height (H) and layer width (W), to control the deposition efficiency and geometric accuracy. These two responses are governed by some independent input variables: arc current (I), arc voltage (V), TS, WFS, working distance, and heat input (HI). The process parameters in the WAAM technique are heavily influenced by the type of material and the design of the fabricated parts. These parameters significantly influence the manufacturing process and are crucial in determining the optimal material properties. Adjusting these WAAM parameters can help achieve the desired material characteristics, which enhances the final product's quality and performance. Figure 2.5 shows typical wall deposition sample built by the WAAM process.

Various types of process parameters involved in the WAAM process can greatly influence both the manufacturing process and the final properties of the components being produced (Dinovitzer et al., 2019). Table 2.1 describes some process parameters.

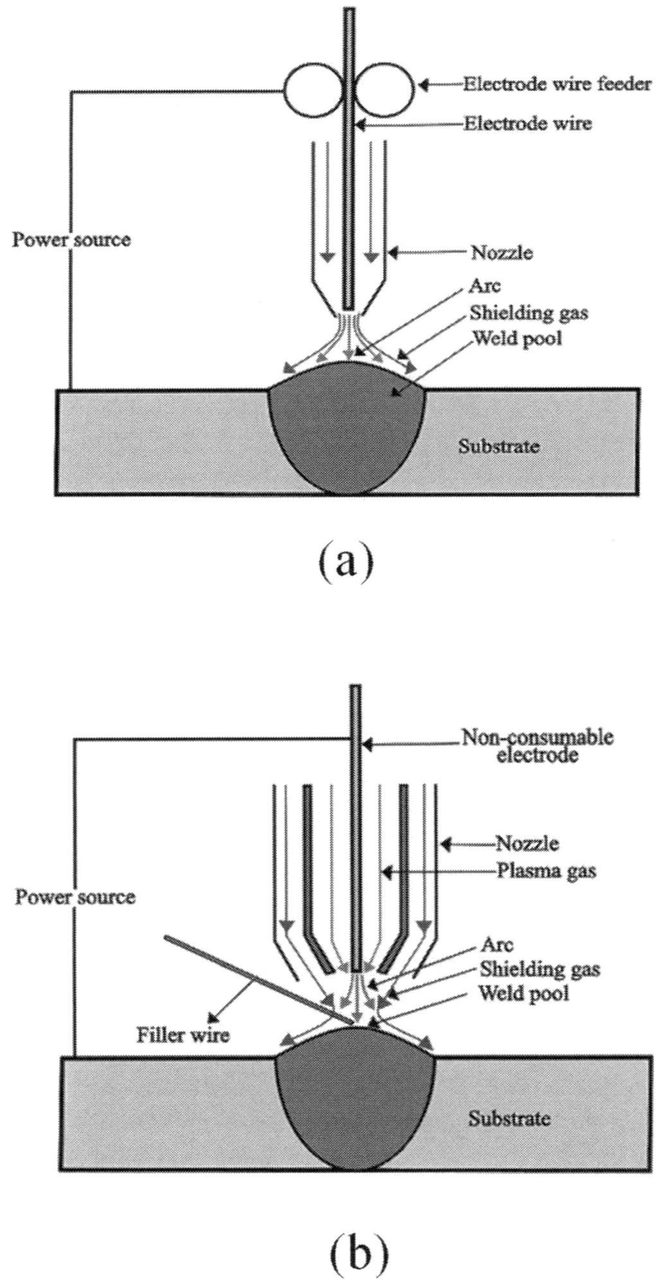

FIGURE 2.4 (a) GMAW, GTAW, and PAW. (b) GTA-DED or GTAW. (c) Plasma arc DED (PA-DED) or PAW.

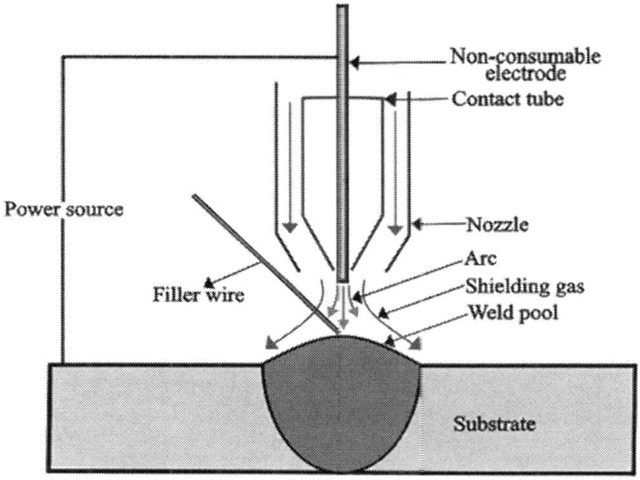

FIGURE 2.4 (Continued).

2.2.4 Optimal Material Characteristics

The optimal material characteristics of WAAM fabricated parts are measured by different factors such as microstructure, microhardness, tensile strength (TS), and surface quality. Several researchers have investigated these factors during other advanced machining processes, too.

2.2.5 Microstructure

Microstructure describes the small-scale structure of a material, which can be observed on a prepared surface using an optical microscope with magnifications exceeding 25×. They are formed during phase changes in a material induced by temperature or pressure shifts, such as when a melt cools and crystallizes into a solid.

An experimental study by Ismail Hemmati et al. (2011) focused on characterizing the microstructure of martensitic stainless steel type 431 through the application of laser-deposited coatings. The team observed and studied the solidification formations depicted in Figure 2.6.

Figure 2.6(a) depicts a cross-section displaying planar growth occurring specifically at the interface between the substrate and the coating. Figure 2.6(b) presents a cellular dendritic structure, showcasing a notable transition from columnar to equiaxed grains observed prominently at the topmost section of the deposited material.

FIGURE 2.5 Typical wall deposition sample fabricated by the WAAM process using 410 martensitic stainless steel as filler material and 304 L austenitic stainless steel as substrate material.

2.2.6 Microhardness

Microhardness is the property of a material that pertains to its capacity to withstand plastic deformation, indentation, penetration, and scratching. The determination of a material's hardness involves applying a Vickers or Knoop indenter with a load ranging from 15 to 1000 gf onto the material's surface.

In addition to their use in microscopic examination, the prepared specimens can be utilized for measuring hardness (Gaikwad et al., 2020). This facilitates the analysis of microhardness, considering the direction of deposition as a result of thermal cycling and heat accumulation. It has the ability to classify diverse varieties of microstructures in the samples. When grain sizes are reduced, the number of grain boundaries increases, and these boundaries serve as barriers to displacements. This relationship is significant because hardness, which measures a material's resistance to plastic deformation, is directly affected by the presence of grain boundaries.

By analyzing the hardness measurements obtained from the test, it may be feasible to estimate the UTS and YS of the deposited material. From the values of the Vickers Hardness (HV) test, the values for UTS and YS can be easily estimated through the correlation explained in.

2.2.7 Tensile Strength

The definition of TS is that when a material is allowed to be stretched or pulled it can withstand the maximum stress it can withstand before it fractures. It is measured

TABLE 2.1
Some Process Parameters Details for WAAM

Process Parameters	Particulars
Travel speed (TS)	The weld bead speed relative to the surface is typically regulated by the travel speed of the welding torch. If the torch travels too slowly, it can cause an excess buildup of heat. Conversely, if the torch moves too rapidly, gaps can form in the weld pool, resulting in a defect known as 'humping'.
Wire feed speed (WFS)	It measures how quickly the wire is fed through the welding torch. It is very important because it could change the shape and dimensions of the weld bead, as well as how much time is required to complete a weld. So, the quality of the weld depends on WFS. The common way to express this is in inches per minute (inch/min).
Current (I) and voltage (V)	Current and voltage are such parameters that are determinants in the form of metal transfer to the piece. Current (I) is related to WFS, as the higher the feed rate the greater the current. So, the rate of deposition is related to the current: the more current is supplied the more quickly fusion will occur. Voltage (V) is adjustable in the power source and is transmitted regularly from the source to the wire. So, if the supplied voltage is greater, the length of the arc will be greater.
Heat input (HI)	Heat input is a very important parameter in the WAAM method. The measure of transferred energy per unit length of weldments is relative. Heat input is a dependent parameter, which can be established by employing the following equation, where 'V' is the arc voltage and 'μ' is the arc efficiency of the process. $\text{Heat Input}(\text{HI}) = \frac{I \times V \times \mu}{TS}$

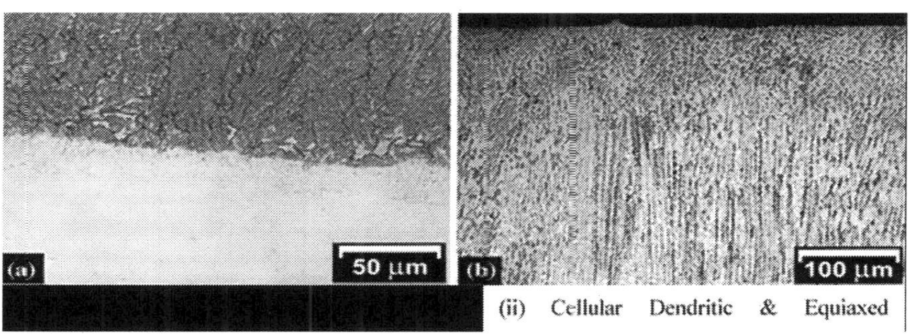

FIGURE 2.6 Solidification microstructures in 431 martensitic stainless steel.

by force per unit cross-sectional area. The transition from elastic to plastic deformation in a material is precisely defined by this specification. Tensile specimens were extracted from two regions in the vertical direction to evaluate the mechanical properties of the component.

The tensile strength of a WAAM part was significantly improved through the implementation of a rolling process, which facilitated the introduction of a high quantity of dislocations and small grains. Recrystallization during a cyclic heating process may release dislocations and strains, but this alone is not enough to counteract the overall impact of rolling. Therefore, the density of dislocations, which is accountable for the enhanced tensile strength, is observable within the interlayer rolled part, leading to a rise in strength.

Tensile tests, a critical aspect of material testing, are commonly performed following the ISO 6892-1:2016 standards. To conform to the specified standard, walls must be constructed with a minimum elevation of 100 mm and a breadth of at least 12 mm. It was achieved by employing a weaving motion in conjunction with the ideal parameters determined during the preliminary deposition investigation, employing a linear motion to create the most substantial walls.

2.3 MATERIALS AND THEIR ALLOYS USED IN WAAM

The WAAM process makes use of a range of materials supplied by select arc welding. These materials include titanium (TiAl6V4B), which is commonly used in the aeronautical industries and industrial chemistry. Inconel 7718 (NiFe19Cr19Nb5Mo3) is another material utilized in the aeronautical industry and for the production of high-temperature tools. Steel (AISI 420, ER316LSi, SMV3S, ER2209 duplex) is employed for assembling and repairing stainless steel parts. Additionally, aluminum (AlSi5, AlSi5GrY, AlMg5Cr, AlSi7MgO, 5Ti, AlCu6MnZrTi) is used for extruded components and foundry repairs. These materials are supplied in wire spools of varying thicknesses.

2.4 RELEVANT LITERATURE ON WAAM

WAAM processes have a remarkable impact on different modern manufacturing-based industries, due to some advantages like high deposition rates, lower production rates, low raw material cost, flexibility concerning the part size, and high efficiency. This process is thus the most significant for the AM process of metallic materials like titanium, aluminum, copper, and steel, according to their requirements and use.

This study examines the principles, process parameters, microstructural characteristics, mechanical properties, heat treatment effects, and defects associated with various applications. Various metals and alloys employed in this process are analyzed based on the findings of various researchers. From the different literacy surveys, it is understood that the entirety of the research can be categorized into four distinct parts: the WAAM hardware system, physical process and analysis, process monitoring, and property characterization. Figure 2.7 represents the outline for the WAAM process review.

2.5 TARGET FUNCTION

The following target functions are documented in the present literature:

Target 1: Influences of WAAM Process Parameters on Weld Bead Geometry
Target 2: Influences of WAAM Process Parameters on Mechanical Properties of Produced Part
Target 3: Influences of WAAM Process Parameters on Microstructure of Produced Part

2.5.1 Influences of WAAM Process Parameters on the Weld Bead Geometry

The process parameters notably influence the shape and size of weld beads in WAAM. The weld beads' characteristics substantially impact geometric correctness and the deposited structure's surface characteristics in both single- and multi-layer deposition processes. It is crucial to manage the WAAM process parameters to accomplish the desired characteristics of bead geometry within the deposited layers. This is crucial for precisely forecasting, managing, and enhancing the surface quality and dimensional correctness of WAAM-manufactured parts. Several research communications have been undertaken to study the influences of WAAM-based process parameters on bead geometry. Figure 2.8(a) shows the deposited weld bead, and Figure 2.8(b) shows the schematic cross-sectional view of the weld bead.

Dinovitzer Malcolm et al. (2021) carried out an experimental study to investigate the influences of varying process parameters in TIG-based WAAM on sample components fabricated using Hastelloy X alloy welding wire, with a 304-grade stainless steel plate used as substrate. To establish the influences of arc current, TS, WFS, and argon flow rate on various responses such as beads' morphology and dimensions, surface roughness of beads, oxidation levels, depth of penetration, and microstructural characteristics, they have applied the Taguchi method and ANOVA. The major influences on the responses were identified by them as the TS and current. They observed that the melt-through depth decreased and roughness increased when TS increased or the current decreased. The three- and eight-layer samples have been analyzed, revealing the presence of three distinct zones within each sample. The matrix grain formations displayed unique features depending on the distribution and size of the molybdenum carbide (MoC and Mo_2C) formations within the zones.

Manokruang (B) Supasit et al. (2020) also carried out experimental research to investigate process parameters involved in the WAAM method and how they impact the geometry development of weld beads. They used ER5356 Al alloy as substrate and a 1.2 mm diameter 5356 Al wire as filler or welding wire.

The heating device was utilized to regulate the temperature of the substrate, with the temperature range spanning from ambient room temperature to a scorching 400°C. This was done by the cold metal transfer (CMT) technology, where modifications were made to the welding parameters, WFS, and TS. Throughout the experiments, a constant ratio λ (WFS/TS) was maintained. The researchers discovered that the geometry of the weld bead was controlled by both the temperature of

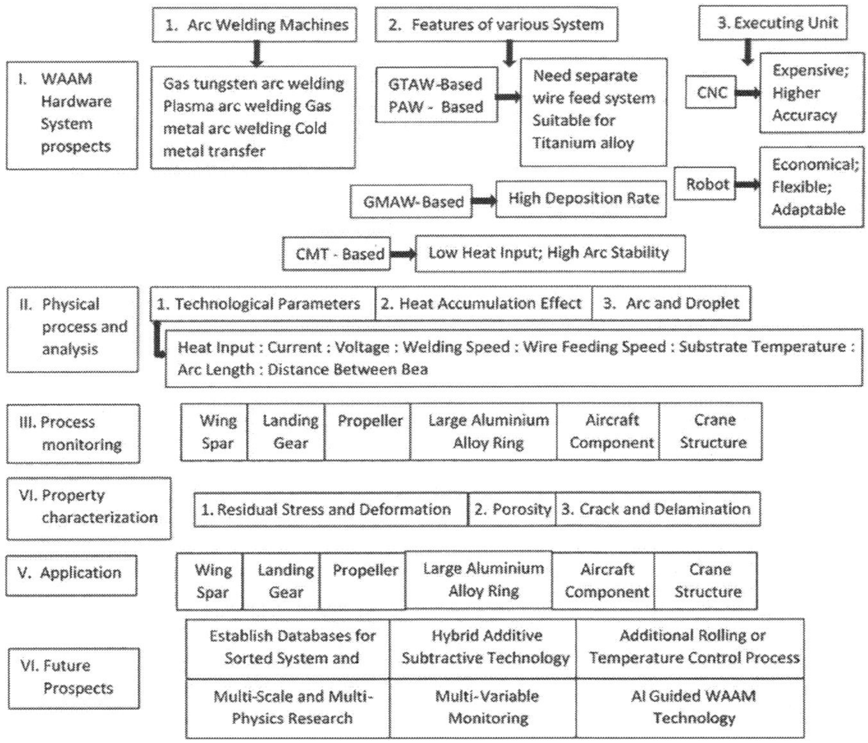

FIGURE 2.7 Outline for WAAM process review.

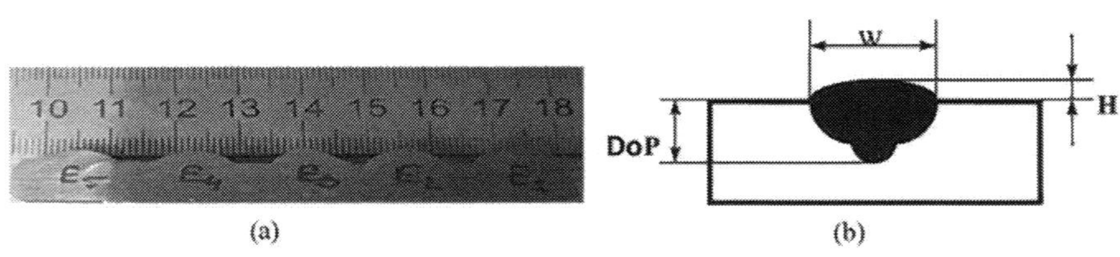

FIGURE 2.8 (a) Deposited weld bead and (b) schematic cross-sectional view of the weld bead and its parameters.

the substrate and the aforementioned parameters. Furthermore, through a series of experiments, it was determined that the key factors driving the geometrical changes were the WFS and the temperature of the substrate (T_{sub}), while keeping the amount of deposited material constant.

An experimental study carried out by Ahsana et al. (2021) to investigate the influence of process parameters on the beads' shape in WAAM of Al0.1CoCrFeNi high-entropy alloy. They used alloy Al0.1CoCrFeNi as substrate and welding wire material. To accomplish the objective of the process, a two-phase investigation was conducted by using pre-alloyed and extruded Al0.1CoCrFeNi wire. During the initial phase, they applied various process parameters like arc current, WFS, and TS to deposit single-layer beads. They have identified an energy density threshold (>80 J/mm³) necessary to achieve uniform and continuous bead formation to measure the bead cross-section by means of monitoring process signals, conducting measurements, and performing statistical analysis.

During the second phase, they applied two different heat input conditions with equal energy concentration to deposit thin-walled, multi-layer structures. Through observations, it was discovered that there were significant columnar grains characterized by cellular-dendritic-substructures and a homogeneous composition. The strength of nearly 260 MPa and UTS of 420 MPa were measured, along with remarkable ductility ranging from 45–55% for both deposits. These properties may surpass those of casting and other AM methods. Enhanced surface quality, greater ductility, and a higher deposition rate are attributed to the increased heat input, which is recognized as the nearly optimal processing condition.

Another experimental study was carried out by Manjhi Shambhu Kumar et al. to investigate the influences of different process parameters like WFS and TS on single tracks of AZ31 Mg metalloids produced through the CMT-WAAM process. They observed that increases in WFS led to greater width, height, depth of penetration, and heat-affected zone in single tracks. In addition, a rise in TS resulted in a decrease in the contact angle and height of the deposited tracks. They used AZ31 Mg metalloids as substrate plates, with a size of $150 \times 50 \times 4$ mm^3 and AZ31 Mg metalloids with 1.2 mm diameter as welding wire or filler material. In this experimental study, they investigated how the bead geometry of deposited tracks was affected by the aforementioned two process parameters: WFS and TS.

After completion of the deposition process, by utilizing the wire-cutting method, they sliced the block into smaller samples to examine the bead size, contact angle, and dilution. Single tracks were separated by the wire electrical discharge machining (EDM) process. They observed that every bead is a single-layer bead, and these beads are produced consistently at a fixed WFS of 6 m min^{-1}, while having different TS values—0.4, 0.5, and 0.6 m min^{-1}, resulting in a bead width that is smaller than its height. They again noticed after increasing the value of WFS to 8 m min^{-1} with setting the value of TS at 0.4, 0.5, and 0.6 m min^{-1} and found that the beads' width and height are appropriate, and due to these the deposited track beads are stable on the substrate plate. By increasing the value of WFS to 10 m min^{-1} along with controlling the values of TS of 0.4, 0.5, and 0.6 m min^{-1}, the bead width expanded, while the height decreased simultaneously, leading to an unstable bead formation.

A study of an experimental nature was conducted by Kumar B. Mohith et al. to recognize the ideal processing parameters in the WAAM of 304 L stainless steel and to assess the influence of input process parameters like arc current (I), arc voltage (V), and TS on the geometrical properties of single weld beads. They used 304 L-stainless steel plates as substrate with the dimensions of $100 \times 80 \times 10$ mm and 304L steel with 0.8 mm diameter as welding wire or filler material. They were designed by the L16 orthogonal array and the Taguchi technique for this experimental study, and three input variables were chosen with four levels for this investigation. They conducted 16 experimental runs of single weld beads and therefore gathered data on the considered responses were those that were taken into account. After performing a series of experiments, they presumed that the prior weld bead run had no impact on the subsequent weld beads' geometrical reactions.

2.5.2 Influences of WAAM Process Parameters on Mechanical Properties of the Produced Part

The choice of process parameters is typically determined by the specific material and the geometry of the produced parts. Regarding the matter of the WAAM process, several parameters like voltage, current, TS, WFS, and flow of shielding gas can influence oxidation & other issues. The material's properties, including mechanical properties, are modified by oxidation, thereby influencing both the process and the resulting properties of the produced parts. Efforts are made to minimize these influences by optimizing the parameters. Several research communications have been undertaken to study the influences of WAAM-based process parameters on the mechanical properties of different WAAMed parts.

Nguyen Van-Thuc et al. (2023) carried out an experimental study to determine the effect of travel rate, voltage, and intensity on the mechanical properties of selected specimens produced by the WAAM technique. They used S20C steel as substrate and 0.8 mm diameter AWS A5.18 ER 70S-6 alloy as welding wire or filler material. They had arranged 12 samples for this experiment, in such a manner that, where each sample number consisted of three samples, along with the WAAM process parameters—travel rate, voltage, and intensity were set up accordingly.

After completion of the deposition process, they cut down the block into smaller parts for testing specimens by the wire EDM technique, and evaluation of specimens is conducted to assess their mechanical properties. The primary objective of the study was to determine the average mechanical properties of WAAM specimens under various process parameters. They found that the values of UTS fall within the range of desired range. When compared with the standard welding wire, they found that their specimens exhibited approximately 21–40% greater strength than that of the original steel wire. Moreover, the WAAM specimens demonstrated superior mechanical properties, showcasing increased UTS and elongation at break values compared to the initial wire.

A study to investigate the influences of arc current, WFS, TS, wire feeding angle, tungsten electrode diameter, flow rate of argon shielding gas, mechanical properties, and wear behavior of high-carbon chromium bearing steel samples produced by the WAAM technique was experimentally studied by Wenjie Zhang (2019). They used GCr18Mo as a substrate and 1.2 mm diameter ER55-Ni alloy material as a welding wire or filler material. After completion of the deposition process, they cut down the block into reduced specimens by the wire cut EDM technique, and the specimens undergo analysis to determine their mechanical properties.

They also observed a gradual decrease in the hardness of the deposition layer in the WAAM specimens, from its highest point to the base. The transition layer was found to have an approximate hardness of 350 HV. Based on the results obtained from the tensile tests, the recorded data indicate that the WAAM specimen shows greater YS and UTS while displaying a lower elongation (EL) in comparison to the specimens of the substrate. The value of YS and UTS of WAAM specimens are recorded at 622 and 998 MPa, respectively, which are 102% and 70% greater than the substrate specimen testing value, while the EL is 20%, representing 46% greater than the substrate specimen.

Zhang Yazhou et al. (2018) also conducted an experimental study to investigate the influences of various process parameters on the mechanical properties of aluminum alloy specimens. They used aluminum alloy 2A12 as substrate and a 1.2 mm diameter $AlCu_6Mn$ alloy material as welding wire or filler material. After completion of the deposition process in VP-CMT arc mode, they cut down the block into smaller specimens using the wire EDM method, and the specimens were investigated to study the mechanical properties.

The outcomes were observed, with process parameters being employed to achieve them, revealing the average value of TS in both orientations of their specimens. Enhanced tensile strength was observed in both the vertical and horizontal specimens when the same parameters were utilized for testing. The range of anisotropy percentage observed varied from 2.3% to 14.9%, with no discernible pattern. The observed anisotropy percentages varied between 2.3% and 14.9%, without any clear pattern.

After the thorough analysis of the sample results, they noticed that in comparison to the vertical direction, it was observed that the WAAMed $AlCu_6Mn$ metalloids displayed significantly improved mechanical properties in the horizontal direction.

2.5.3 Influences of WAAM Process Parameters on Microstructure of Produced Part

Process parameters usually depend on the used material and the geometry of the manufactured parts. For the WAAM process, different types of parameters like voltage, current, TS, WFS, and shielding gas flow can affect oxidation and other complications. Oxidation impacts the characteristics of the material, such as its microstructure, which in turn affects the manufacturing process and the properties of the final components. The optimization of parameters aims to reduce these effects.

The observation reveals that an increase in heat input during the WAAM process leads to a more pronounced microstructural characteristic characterized by fewer grain boundaries. This, in turn, encourages the occurrence of liquation along the grain boundaries and the entrapment of gas. Multiple research investigations have been undertaken to study the influences of WAAM-based process parameters on the microstructural characteristics of different WAAMed parts.

Su Chuanchu et al. (2019) conducted an experimental study to investigate the effects of different heat inputs by adjusting different process parameters, that is, WFS and TS, on Al-Mg metalloid microstructure made up by the WAAM process. They used a 6061-T6 aluminum alloy plate as a substrate and 1.2 mm diameter ER5356 alloy material as a welding wire or filler material.

After completion of the deposition process based on the CMT technology, they cut down the block by utilizing the wire-cutting method into smaller specimens, and the specimens are analyzed for microstructural studies. They observed that grain size fluctuations in the inner-layer region (NLR) ranged from 42.9 to 88.7 μm, while in the interlayer region (TLR) they ranged from 37.7 to 77.6 μm, when the WFS was 7.0 $m.min^{-1}$ and TS was 0.9 $m.min^{-1}$. Conversely, larger grains were found in substantial quantities in these regions under different combinations of WFS/TS.

Dinovitzer Malcolm and Aidun D.K. also carried out an experimental study of empirical investigation to examine the influences of various process parameters, that is, arc current, arc voltage, TS, shielding gas, interlayer idle time (IIT), deposition strategy, number of layers, WFS, and processing time regarding the microstructure of Inconel 718 (a Nickel based super alloy) samples produced by the WAAM technique. They used A36 steel as substrate and $ERNiFeCr_2$ Inconel 718 with a 0.9 mm diameter as welding wire or filler material. The mentioned WAAM process parameters were set up accordingly. From the parametric research, they noticed that, the deposition process each containing of 30 layers led to the formation of one tube and four linear thin wall components.

In order to conduct the microstructural assessment, they opted to section the dominant portions of the walls and tube along the layer direction. They sectioned the specimens by the water jet cutting device and specimens are further analyzed for microstructural studies. They observed the microstructure by scanning electron microscopy.

After completing the deposition process, an analysis was conducted to assess the impact of various process parameters on solidification morphology. Samples were meticulously examined to evaluate the microstructural properties of longitudinal sections. It was noted that the TLR exhibited a finer solidification morphology compared to the inner-layer zone. In all WAAM products, the typical columnar solidification morphology appeared coarser in the inner layer of the tube because of lower cooling rates throughout the continuous deposition process, resulting in an IIT of zero. Comparison between the NLR and TLR of WAAM components showed nearly identical microstructures, suggesting minimal influence of processing parameters such as shielding gas type, IIT variation (20 to 60 s), and deposition strategy on microstructural development.

A research study was performed experimentally by Ty et al. (2022) to investigate to gain insights into the effects of various process parameters like TS, WFS, and current (I); it is crucial to examine the capabilities of PAW as a heat source in DED-W on the microstructure of Ti-6Al-4V titanium (Ti) alloys fabricated by the WAAM process. They used an annealed Ti-6Al-4V grade 5 alloy plate (thickness of 30 mm) as substrate and 1.2 mm diameter 23-Grade Ti-6Al-4V-ELI alloy material as welding wire or filler material.

After completion of the deposition process based on PAW technology, they cut down the block by utilizing the wire-cutting method into smaller specimens, and the specimens are analyzed for their microstructural studies (Gaikwad et al., 2020). They have identified the pause between layers as a significant process parameter that governs the macro- and microstructure. The observed thicker plates in this experiment, unlike those in existing literature, are likely due to variations in the process parameters, with the pause between layers mentioned earlier. Moreover, the expanded top section indicates the significance of the walls above the β transit possesses a substantial volume, matching the β grain size from before, irrespective of the deposition parameters.

2.6 CONCLUSION

Based on the study and analysis of this comprehensive review, the following conclusions can be drawn:

i. The various process parameters have a significant impact on the characteristics of the weld bead.
ii. The variations in the solidification method of the metalloid were brought about by the WAAM, leading to changes in its phases.
iii. WAAMed metallic alloys demonstrate significantly improved mechanical properties when oriented horizontally rather than vertically, with the level of anisotropy fluctuating unpredictably.
iv. Discrepancies in the microstructure of the sectors arise from distinct temperature variations and heat exposure encountered by individual deposited beads. Specifically, the metal in the upper segment undergoes a more pronounced temperature decrease because of the variance in the air and lower-layer temperature.
v. The ductility of the WAAM-manufactured parts was observed to be lower compared to the wrought materials/alloys, particularly in terms of mechanical characteristics. The inter-dendrite area of the WAAM products showed a higher density of the laves phase, leading to a decrease in strength along the layer direction.

ACKNOWLEDGEMENTS

The authors wish to express their gratitude to the Director of CSIR-CMERI for giving permission to carry out the research work at CSIR-CMERI, Durgapur, which enabled the successful execution of the research endeavors. They also express their sincere thanks to the Chancellor, SRMU, for creating a research-oriented environment at the SRMU campus and the Director, NSCD (a unit of NCSM), Ministry of Culture, Government of India, for continuous support and necessary guidance to carry out the research work.

REFERENCES

Ahsana, M. R., Seo, G.J., Fan, X., Liaw, P. K., Motaman S., Haase, C., & Kim, D. B. (2021). Effects of Process Parameters on Bead Shape, Microstructure, and Mechanical Properties in Wire + Arc Additive Manufacturing of Al0.1CoCrFeNi High-Entropy Alloy. Journal of Manufacturing Processes, 68, 1314–1327.

Ding, D., Pan, Z., Dominic, C., & Li, H. (2015). Wire-Feed Additive Manufacturing of Metal Components: Technologies, Developments and Future Interest. The International Journal of Advanced Manufacturing Technology, 81(1–4), 465–481. https://doi.org/10.1007/s00170-015-7077-3

Dinovitzer, M., & Aidun, D. K. (2021). Effects of Processing Parameters on Wire Arc Additive Manufactured Inconel® 718.

Dinovitzer, M., Chen, X., Laliberte, J., Huang, X., & Frei, H. (2019). Effect of Wire and Arc Additive Manufacturing (WAAM) Process Parameters on Bead Geometry and Microstructure. Additive Manufacturing, 26, 138–146.

Fidan, I., Alshaikh Ali, M., Naikwadi, V., Gudavasov, S., Mahmudov, M., Mohammadizadeh, M., . . . & Sharma, A. (2024). Nano-Level Additive Manufacturing: Condensed Review of Processes, Materials, and Industrial Applications. Technologies, 12(7), 117.

Gaikwad, M. U., Gaikwad, P. U., Ambhore, N., Sharma, A., & Bhosale, S. S. (2024). Powder Bed Additive Manufacturing Using Machine Learning Algorithms for Multidisciplinary Applications: A Review and Outlook. Recent Patents on Mechanical Engineering, 18(1), 12–25.

Gaikwad, M. U., & Jatti, V. S. (2020). Estimation of Surface Integrity Parameters in Electrical Discharge Machining (EDM) Process-a Review. In Techno-Societal 2018: Proceedings of the 2nd International Conference on Advanced Technologies for Societal Applications (Volume 2, pp. 601–614). Springer International Publishing. https://doi.org/10.1007/978-3-030-16962-6_61

Gaikwad, M. U., Krishnamoorthy, A., & Jatti, V. S. (2020). Investigation on Effect of Process Parameter on Surface Integrity during Electrical Discharge Machining of NiTi 60. Multidiscipline Modeling in Materials and Structures, 16(6), 1385–1394. https://doi.org/10.1108/MMMS-10-2019-0179

Hemmati, I., Ocelík, V., & De Hosson, J. Th. M. (2011). Microstructural Characterization of AISI 431 Martensitic Stainless-Steel Laser-Deposited Coatings. Journal of Materials Science, 46, 3405–3414.

Li, Y., Su, C., & Zhu, J. (2016). A Comprehensive Review of Wire Arc Additive Manufacturing: Hardware System, Physical Process, Monitoring, Property Characterization, Application and Future Prospects. Results in Engineering, 13, 100330.

Manjhi, S. K., Sekar, P., Bontha, S., & Balan, A. S. S. (2013). Effect of CMT-WAAM Process Parameters on Bead Geometry, Microstructure and Mechanical Propertiesof AZ31 Mg Alloy.

Manokruang (B), S., Vignat, F., Museau, M., & Limousin, M. (2020, June). Process Parameters Effect on Weld Beads Geometry Deposited by Wire and Arc Additive Manufacturing (WAAM). In International Joint Conference on Mechanics, Design Engineering & Advanced Manufacturing (pp. 9–14). Cham: Springer International Publishing.

Mohith Kumar, B., Sandeep, K., Johnson, K., Sai, M., & Mr. Sreenivasa Reddy, K. (2024). Optimization of Process Parameters of Wire Arc Additive Manufacturing of 304L Stainless Steel Parts Using Grey Analysis Method.

Nguyen, V.-T., Minh, P. S., Uyen, T. M. T., Do, T. T., Ngoc, H. V. T., Le, M.-T., & Tien Nguyen, van T. (2023). WAAM Technique: Process Parameters Affecting the Mechanical Properties and Microstructures of Low-Carbon Steel.

Sharma, A., Babbar, A., Tian, Y., Pathri, B. P., Gupta, M., & Singh, R. (2023). Machining of Ceramic Materials: A State of the Art Review. International Journal on Interactive Design and Manufacturing (IJIDeM), 17(6), 2891–2911.

Sharma, A., Jain, V., & Gupta, D. (2018). Characterization of Chipping and Tool Wear during Drilling of Float Glass Using Rotary Ultrasonic Machining. Measurement, 128, 254–263.

Sharma, A., Jain, V., & Gupta, D. (2019a). Comparative Analysis of Chipping Mechanics of Float Glass during Rotary Ultrasonic Drilling and Conventional Drilling: For Multi-Shaped Tools. Machining Science and Technology, 23, 547–568.

Sharma, A., Jain, V., & Gupta, D. (2019b). Multi-Shaped Tool Wear Study during Rotary Ultrasonic Drilling and Conventional Drilling for Amorphous Solid. Proceedings of the Institution of Mechanical Engineers, Part E: Journal of Process Mechanical Engineering, 233, 551–560.

Sharma, A., Jain, V., & Gupta, D. (2021). Effect of Pre and Post Tempering on Hole Quality of Float Glass Specimen: For Rotary Ultrasonic and Conventional Drilling. Silicon, 13, 2029–2039.

Sharma, A., Jain, V., & Gupta, D. (2022). Mathematical Approach on Chipping Volume Estimation Generated during Rotary Ultrasonic Drilling for Float Glass. Proceedings of the National Academy of Sciences, India Section A: Physical Sciences, 92, 285–291.

Sharma, A., Jain, V., Gupta, D., & Babbar, A. (2020). A Review Study on Miniaturization. In Advanced Manufacturing and Processing Technology (First edition, pp. 111–131). Boca Raton, FL: CRC Press.

Su, C., Chen, X., Gao, C., & Wang, Y. (2019). Effect of Heat Input on Microstructure and Mechanical Properties of Al-Mg Alloys Fabricated by WAAM. Applied Surface Science, 486, 431–440.

Tripathi, N. M., & Sharma, A. (Eds.). (2024). Advances in Pre-and Post-Additive Manufacturing Processes: Innovations and Applications. CRC Press.

Ty, A., Balcaen, Y., Mokhtari, M., & Alexis, J. (2022). Influence of Deposit and Process Parameters on Microstructure and Mechanical Properties of Ti6Al4V Obtained by DED-W (PAW). Journal of Materials Research and Technology, 18, 2853–2869.

Zhang, W., Lei, W., Zhang, Y., & Liu, X. (2019). Microstructure and Mechanical Properties of Wire Arc Additive-Manufacturing High-Carbon Chromium Bearing Steel.

Zhang, Y., Gao, M., & Zeng, X. (2018). Effect of Process Parameters on Mechanical Properties of Wire and Arc Additive-Manufactured AlCu6Mn.

3 Teaching Learning–Based Optimization for Optimizing Process Parameter in Wire Electrical Discharge Machining (WEDM)

Shalaka Kulkarni, Omkar K. Kulkarni, Samidha A. Jawade, Swanand Pachpore, and Ganesh Kakandikar

3.1 INTRODUCTION

With the continuous advancement of the machinery industry, there has been a noticeable uptick in the demand for alloy materials that exhibit exceptional qualities such as toughness, hardness, and impact resistance. This need for high-performance materials has spurred the utilization of wire EDM, a cutting-edge technology that can effectively process materials of varying hardness levels. What sets wire EDM apart is its capability to handle complex shapes and contours with precision, a feat that poses challenges for conventional cutting methods. These intricate designs are often seen in the production of items like complex punches, dies, and various spindles. Among the materials widely employed in this manufacturing landscape is high-carbon and high-chromium steel, appreciated for its superior machinability, wear resistance, compressive strength, and corrosion resistance. As industries evolve and production requirements become more intricate, the importance of leveraging innovative solutions like wire EDM technology becomes increasingly evident (Kumar et al., 2024). This dynamic environment calls for materials that not only meet but exceed the expectations of modern manufacturing demands, underscoring the indispensable role played by alloy materials with exceptional properties in driving industrial progress.

In the wire electrical discharge machining (WEDM) process, as also emphasized by Sharma et al. (2024), a typical thermoelectric method is employed, where the

 DOI: 10.1201/9781003502647-4

material removal from the workpiece is facilitated through the continuous generation of sparks between the tool and the workpiece. Throughout this procedure, deionized water acts as the dielectric medium, playing a crucial role in both insulating and ionizing the system. These sparks are responsible for producing substantial amounts of energy, resulting in the tool and workpiece being heated up significantly. However, the dielectric medium efficiently absorbs and dissipates this heat, preventing any overheating issues. Moreover, the dielectric flushing in the spark gap serves the dual purpose of anticipating and mitigating debris contamination and early discharges. Furthermore, this flushing mechanism is pivotal in regulating the workpiece's temperature, offering a means to alter its properties as needed. By ensuring proper cooling through dielectric flushing, the adverse effects of thermal expansion within the workpiece are mitigated. Failure to implement this cooling system would give rise to challenges related to position accuracy and workpiece dimensions due to the uncontrollable thermal expansion dynamics within the material being processed.

The proposed methodology, which integrates two cutting-edge optimization technologies, offers a comprehensive approach to enhancing process efficiency. Initially, Taguchi's design of experiments (DOEs) is employed to meticulously examine the impact of various process parameters on surface roughness (SR). Through the application of linear regression analysis, a robust mathematical model is constructed to establish correlations between SR and the identified process parameters. This model serves as a valuable tool in predicting the SR values based on different process conditions, thereby enabling a more precise control over the manufacturing process (Sharma et al., 2020a, 2020b, 2024a, 2024b; Babbar et al., 2021; Bhayana et al., 2023).

Furthermore, to fine-tune the optimization process, the teaching learning–based optimization (TLBO) algorithm is implemented to identify the optimal solution with maximum efficiency. This algorithm harnesses the principles of teaching and learning to navigate the parameter space and converge on the most favorable outcome. By tapping into the collective intelligence and adaptability of this optimization approach, the methodology ensures that the process is fine-tuned to achieve the desired results efficiently and effectively.

In essence, the integration of Taguchi's DOE and TLBO algorithm in the proposed methodology not only empowers manufacturers to gain deeper insights into the process parameters' impact on SR but also equips them with a powerful optimization tool to enhance the overall performance of the manufacturing process. Through this holistic and data-driven approach, businesses can leverage cutting-edge technologies to drive continuous improvement and achieve superior quality outcomes in their manufacturing operations. Various evolutionary and bioinspired optimization algorithms are available these days, which include particle swarm optimization (Patnaik & Mahapatra, 2006), genetic algorithm (Kulkarni & Rodge, 2014), cohort intelligence algorithm (Kulkarni et al., 2015), ant colony optimization (Bhoskar et al., 2015), gray wolf algorithm (Sharma et al., 2015), grasshopper algorithm (Neve et al., 2017), and salp swarm algorithm (Mhatugade et al., 2019). The TLBO algorithm (Rao & Patel, 2013) has the unique advantages and functions described in the next section.

3.2 MAJOR INFLUENCING PARAMETERS

In the realm of WEDM, performance evaluation hinges on a multitude of response variables. These variables serve as indicators of the machining quality achieved during the process. Surface finish, cut width (cut), and material removal rate (MRR) stand out among the prominent aspects that are scrutinized to gauge the effectiveness of WEDM operations. Each of these variables plays a crucial role in providing insights into the precision, efficiency, and overall quality of the machined parts. Moreover, it's essential to highlight that these key response variables not only act as performance metrics but also are intricately intertwined with various processing parameters. These parameters encompass a wide array of factors, ranging from discharge current to pulse duration, pulse frequency, line speed, line tension, and dielectric flow. The interplay between these parameters and the response variables ultimately determines the outcome of the WEDM process.

By meticulously adjusting and optimizing these processing parameters, such as wire tension, machining current, and servo voltage (SV), manufacturers can effectively fine-tune the WEDM setup to achieve the desired outcomes in terms of surface finish, cut width, and MRR. This level of detailed adjustment and optimization is essential in the intricate art of WEDM. By comprehensively understanding the complex relationship between the response variables and processing parameters, operators gain the knowledge and insight required to make informed decisions that invariably result in elevated machining performance and superior part quality. Such profound understanding and skill allow operators to navigate the delicate balance required in the performance of WEDM processes, where the manipulation of key response variables—like pulse-on time and flushing pressure—is as vital as optimizing crucial processing parameters such as wire speed and electrode gap. This balance, akin to a finely tuned instrument, is the crux of achieving consistency and high-quality results in the sphere of precision machining, where every adjustment and decision plays a role in shaping the final product. Mastering this balance is pivotal not only in achieving the desired outcomes efficiently but also in setting the stage for innovation and advancement in the ever-evolving field of modern manufacturing.

3.2.1 MRR

In the process of WEDM, a notable phenomenon occurs as a series of cutting sparks form between the tool and the workpiece, gradually causing corrosion within the material of the workpiece itself. To facilitate this intricate process, it is crucial to submerge both the tool electrode and the workpiece in a specialized liquid dielectric solution. Within this dielectric medium, the spark generated between the tool and workpiece serves to heat and melt the material, thereby facilitating the evaporation of minute traces of material from the workpiece surface. This dissolved material is then effectively drained and washed away using the dielectric fluid. Moreover, a critical aspect to consider in this process is the importance of parameters such as high pulse-on time (TON) and peak current, which significantly impact the MRR. These parameters directly influence the efficiency and productivity of the WEDM process, with higher values resulting in an increased MRR. When TON is high, the duration for which the current is applied to create the spark is longer, leading to more material being melted and removed in each cycle. Similarly, higher peak current increases the intensity of the spark, thereby enhancing the MRR. However, it is essential to also note that a balance must be maintained with factors like off time (TOFF) and SV, as these also play a crucial role in determining the overall effectiveness and quality of the machining operation. TOFF, time interval between successive sparks, allows the dielectric fluid to cool and flush away the eroded material, preventing excessive heating and potential damage to the workpiece. Adequate TOFF is necessary to ensure that the machined surface remains smooth and free of defects. Similarly, the SV controls the gap between the tool and the workpiece, influencing the stability and uniformity of the spark generation. Proper adjustment of SV ensures consistent machining performance and quality. Thus, by carefully adjusting and optimizing these key parameters, manufacturers can achieve the desired MRRs while ensuring precise and controlled machining outcomes. Optimal settings for TON, TOFF, peak current, and SV are essential for balancing high productivity with high-quality finishes in the WEDM process. Through empirical testing and advanced optimization algorithms, such as the Taguchi method or TLBO, it is possible to determine the ideal parameter settings that maximize MRR and minimize surface defects, ultimately enhancing the efficiency and precision of WEDM operations (Rao & Venkaiah, 2013; Saklecha et al., 2022).

3.3 PROPOSED METHODOLOGY

The experiments in this study were meticulously designed using Taguchi's DOEs, which involved considering six input parameters with the SR performance measure. An in-depth analysis was then conducted to assess the impact of these input parameters on SR, ultimately pinpointing the parameter that exerted a significant effect. This phase of the study ensured a systematic examination of each parameter's influence, leading to a clearer understanding of their individual and combined effects on SR. Subsequently, a linear regression was performed to craft a mathematical model, which provided a quantitative representation of the relationship between the input parameters and the SR. This model served as a foundational tool for predicting SR outcomes based on various parameter settings. After establishing the regression model, the salp swarm optimization (SSA) algorithm was employed to pinpoint the optimal solution as part of further problem-solving efforts (Tiwari et al., 2022). The SSA, known for its efficiency in finding global optima in complex search spaces, was instrumental in refining the process parameters to achieve the lowest possible SR. The selection of process parameters was undertaken via an exhaustive review of existing literature and the implementation of pilot experimentation to fine-tune the parameters. This rigorous approach ensured that the parameters chosen for the study were both relevant and optimal for the specific materials and conditions being tested. To enhance the depth of the analysis, the results were thoroughly compared with the insights derived from Taguchi's signal-to-noise (S/N) ratio analysis. The S/N ratio analysis provided a robust framework for evaluating the performance of different parameter settings under varying conditions, helping to identify settings that consistently produced low SR despite noise factors. This comprehensive approach ensured a robust evaluation of the various input parameters and their impact on the SR, leading to a more precise and informed decision-making process. By integrating the Taguchi method, linear regression, and SSA, the study effectively combined statistical rigor with advanced optimization techniques. This integration not only identified the most influential parameters but also fine-tuned them to achieve optimal SR, demonstrating the effectiveness of a multifaceted analytical strategy in solving complex engineering problems. The results of this study contribute to the broader understanding of process optimization and provide valuable insights for industrial applications involving high-carbon and high-chromium steels.

3.4 DESIGN OF EXPERIMENTS

The DOE technique stands as an indispensable asset for designers seeking to enhance their design processes (Gaikwad et al., 2020). Rather than solely focusing on isolated input parameters, which can often overlook significant interactions, DOE empowers designers to comprehensively evaluate the effects of these variables. By delving into the intricate dynamics at play, designers can effectively unravel the complexities within a system and optimize the output response. Through the systematic application of DOE, designers acquire a refined understanding of how various factors influence one another, paving the way for the creation of designs that exhibit remarkable resilience and efficiency. This informed approach not only refines the design process but also fuels innovation by uncovering novel insights that may have otherwise gone unnoticed. It is this emphasis on thorough analysis and cross-factor examination that catalyzes the evolution of standard designs into highly robust, high-performing solutions that surpass industry standards.

Moreover, DOE transcends traditional design methodologies by acknowledging and leveraging the nuanced relationships between different design components. This holistic perspective instills a mindset of adaptability and foresight, enabling designers to anticipate challenges and proactively address them. Ultimately, the adoption of DOE represents a paradigm shift in design thinking—one that champions not only excellence in individual components but also the harmonious synergy that underpins exceptional design outcomes.

Moreover, DOE plays a pivotal role in the identification of critical components and vulnerable areas within a design that may be prone to issues or failures (Mhatugade et al., 2019). By leveraging DOE, designers can pinpoint potential weaknesses and challenges before entering the actual production phase, allowing for the optimization of the design process and overall outcomes. Through the systematic evaluation of sensitive parts and areas, designers can proactively address potential problems, thus enhancing the overall quality and reliability of the final product. In essence, the effective utilization of DOE empowers designers to preemptively detect and mitigate design flaws, facilitating the creation of innovative and resilient solutions that align with desired performance objectives (Patnaik & Mahapatra, 2006)

3.4.1 Selection of Orthogonal Array

The Taguchi method stands as a widely acclaimed approach utilized in the DOEs. It was revolutionary when a Japanese scientist named Genichi Taguchi first introduced

TABLE 3.1
Control Factors and Levels

Factors	Level 1	Level 2	Level 3	Level 4	Level 5	UNITS
WT	500	600	700	800	900	kg
WF	5	6	7	8	9	mm/sec
TOFF	3	4	5	6	7	μsecond
TON	4	5	6	7	8	μsecond
LF	5	6	7	8	9	kg/cm^2
UF	6	7	8	9	10	kg/cm^2

the concept of the orthogonal array of experiments. This ingenious technique has found its applications across various fields of engineering, playing a crucial role in optimizing process parameters. By leveraging the Taguchi method, researchers and engineers can significantly boost the efficiency of information generation, transcending concerns such as usage conditions, manufacturing variations, and part deterioration (Tupe et al., 2017a, 2017b; Saklecha et al., 2023). This, in turn, translates to notable reductions in research and development costs. Not only does the method lead to remarkable reductions in development time, but it also aids in the identification of pivotal design parameters that directly impact operations, performance, and expenditure. Making informed decisions about process parameters can substantially enhance tolerance levels, paving the way for the efficient utilization of low-cost components and production systems (Reddy et al., 2012; Burande et al., 2021). Through these strategic enhancements, manufacturing and operational costs can be dramatically decreased, thereby fostering cost-effective practices in production processes. The focus on key input process parameters like pulse-on time (TON), pulse-off time (TOFF), wire feed (WF), wire tension (WT), overshoot (UF), and undershoot (LF) underscores the precision ingrained in such studies. Each parameter is meticulously categorized into five proficiency levels, a testament to the methodical approach upheld in the research. The deliberate selection of the L25 orthogonal sequence (OA) underscores the meticulous planning that underpins the allocation of control factors and their respective levels. Detailed illustrations in Table 3.1 elucidate the variegated levels of control factors, while Table 3.2 provides a comprehensive overview of the experimental design, complete with specified values, anchoring the study in concrete and measurable data (Kulkarni & Kulkarni, 2018).

3.4.2 Single-Objective Optimization

With the aid of the S/N ratio, it becomes straightforward to identify the optimal combination of various process parameters. This critical aspect involves a detailed analysis of the response variable (SR) utilizing the S/N ratio methodology. The investigation reveals that a specific set of process parameters indeed yields the most favorable outcome for the response variable, denoted as SR (Nene & Kulkarni, 2022; Kulkarni & Kakandikar, 2023a). Drawing inspiration from Taguchi's methodology, a

TABLE 3.2
L25 OA with Their Levels, SR, and S/N Ratio

Exp. No	WT	WF	TOFF	TON	LF	UF	SR	S/N Ratio for SR
1	500	5	3	4	5	6	2.513	−8.0039
2	600	6	4	4	6	7	1.841	−5.3011
3	700	7	5	4	7	8	1.771	−4.9644
4	800	8	6	4	8	9	1.549	−3.8011
5	900	9	7	4	9	10	1.519	−3.6312
6	900	8	3	5	7	7	1.488	−3.4521
7	500	9	4	5	8	8	1.706	−4.6396
8	600	5	5	5	9	9	1.656	−4.3813
9	700	6	6	5	5	10	1.516	−3.614
10	800	7	7	5	6	6	1.681	−4.5114
11	800	6	3	6	9	8	2.923	−9.3166
12	900	7	4	6	5	9	1.903	−5.5888
13	500	8	5	6	6	10	1.683	−4.5217
14	600	9	6	6	7	6	1.69	−4.5578
15	700	5	7	6	8	7	1.665	−4.4283
16	700	9	3	7	6	9	1.718	−4.7005
17	800	5	4	7	7	10	1.683	−4.5217
18	900	6	5	7	8	6	1.6	−4.0824
19	500	7	6	7	9	7	1.911	−5.6253
20	600	8	7	7	5	8	1.846	−5.3247
21	600	7	3	8	8	10	1.731	−4.766
22	700	8	4	8	9	6	1.763	−4.9251
23	800	9	5	8	5	7	1.88	−5.4832
24	900	5	6	8	6	8	1.559	−3.857
25	500	6	7	8	7	9	1.81	−5.1536

loss function is meticulously crafted to measure the disparity between the anticipated value and the actual experimental results. This loss function is then transformed into an S/N ratio metric, further streamlining the evaluation process. Notably, the quality attributes within the S/N ratio analysis fall under three distinct categories: the smaller the better, nominally better, and the larger the better (Kumar & Singh, n.d.; Khare et al., 2022). By leveraging S/N analysis, it becomes possible to assess the S/N ratio for every level of process parameters, wherein a higher ratio signifies superior quality attributes, irrespective of their designated category (Bogar et al., 2023; Gaikwad & Jatti, 2020, Gaikwad et al., 2023). Consequently, determining the optimal level of process parameters entails selecting the configuration yielding the highest S/N ratio (Sharma et al., 2015; Patel et al., 2022). To elucidate, in the "smaller-the-better" category, the S/N ratio is calculated to minimize the variance around the target value, reducing undesired variability. Conversely, in the "larger-the-better" category, the S/N ratio aims to maximize the response variable's magnitude. For the "nominally better" category, the goal is to achieve a response as close as possible to a specified target value, minimizing deviations.

Subsequently, this knowledge can be harnessed to forecast the most efficient combination of process parameters. To culminate the extensive process optimization journey, a confirmation experiment is executed to validate the effectiveness of the identified optimal process parameters within the experimental design domain (Kakandikar et al., 2020a; Gaikwad & Jatti, 2020). This step is crucial as it ensures that the theoretical optimal settings translate effectively into practical, real-world scenarios. The loss function of SR is defined by Eq. (3.1), which serves as the foundation for converting the variability in the response variable into a quantifiable metric. This equation typically incorporates the mean and variance of the response variable, facilitating a comprehensive evaluation of performance. By systematically applying the S/N ratio analysis, researchers can derive a robust understanding of the impact of each process parameter and its optimal setting. This methodological rigor ensures that the final selection of process parameters not only enhances the quality of the SR but also improves the overall efficiency and reliability of the machining process. The integration of empirical data, statistical analysis, and optimization algorithms, as demonstrated in this study, underscores the significance of a multifaceted approach to achieving superior machining outcomes.

$$L_{SB} = \frac{1}{n}\sum_{i=1}^{n} Y_{sr}^2 \tag{3.1}$$

3.4.3 Influences on SR

The delta values of WT, WF, TOFF, TON, LF, and UF are 1.259, 1.41, 1.757, 1.563, 1.467, and 1.089, respectively, presented in Table 3.3.

The case study featuring SR as a primary subject serves as a compelling and illustrative example of the fundamental principle that often "smaller is better." It strikingly demonstrates how even the most minor adjustments in a range of interrelated parameters can lead to profound and impactful changes in the overall outcome. By conducting a detailed and comprehensive examination of the extensive data set outlined in the table, it becomes increasingly evident that the variable known as TOFF exerts an undeniable influence on the performance of SR, essentially steering

TABLE 3.3
Response for S/N Rations Smaller Is better (SR)

Level	WT	WF	TOFF	TON	LF	UF
1	−5.603	−5.216	−6.048	−5.14	−5.589	−5.038
2	−4.578	−4.858	−4.995	−4.12	−4.866	−5.494
3	−4.53	−5.62	−4.687	−5.683	−4.526	−5.091
4	−4.343	−4.725	−4.291	−4.851	−5.527	−4.405
5	−5.576	−4.211	−4.61	−4.837	−4.122	−4.602
Delta	1.259	1.41	1.757	1.563	1.467	1.089
Rank	5	4	1	2	3	6

its trajectory through the welding process with a notable degree of authority and control. In contrast, UF emerges as a factor that, while still crucial, plays a relatively less definitive role in determining the ultimate success and efficacy of the SR welding procedures. Apart from these primary influential factors, a myriad of additional variables collectively contribute to the nuanced and intricate characteristics that define the SR welding process. Delving deeper into this complex landscape, the precise timing intricacies associated with pulse-on time (TON) are singled out as a key and indispensable driving force that significantly shapes the final quality of the SR output. Furthermore, elements like lower flush (LF) introduce their own distinct and unique nuances into the elaborate tapestry of interconnected variables that collectively impact the SR welding outcomes. Beyond these elements, factors such as wire feed (WF) and wire tension (WT) come into play as additional contributors, further underscoring the multifaceted nature and complexity of influences that combine to shape the intricate and detailed process of SR welding. It is truly the intricate and harmonious interplay of all these diverse parameters that forms the intricate matrix of influences that converge to define the overall impact on SR welding, thereby highlighting the intricate, detailed, and multifaceted nature of this specialized welding technique.

3.5 PROBLEM FORMULATION

Regression analysis, a statistical technique, plays a key role in determining and quantifying the relationship between a dependent variable, also known as the response variable, and independent variables, which are the input control factors (Kulkarni & Kakandikar, 2022). This analytical process is commonly conducted using specialized software such as MINITAB 17, which aids in calculating and interpreting the results effectively. Within the context of Eq. (3.2), specifically focused on a linear regression equation involving six input control factors and a variable denoted as SR, the aim is to establish a predictive model that captures the influence and impact of each input factor on the overall relationship with the dependent variable. By examining the coefficients and significance levels within the regression equation, researchers can gain valuable insights into how these factors collectively contribute to the variation in the response variable. For instance, a significant positive coefficient for a particular factor indicates that as this factor increases, the dependent variable tends to increase as well. Conversely, a significant negative coefficient suggests an inverse relationship.

Furthermore, the regression equation serves as a mathematical representation of the underlying relationships, allowing for the identification of patterns, trends, and potential correlations among the variables under study. By plotting the residuals and examining goodness-of-fit measures such as R^2 and adjusted R^2, researchers can assess the model's accuracy and predictive power. These diagnostics help determine whether the model appropriately captures the data's variability and whether any assumptions of linear regression are violated. Through this comprehensive analysis, it becomes possible to make informed decisions and draw meaningful conclusions based on the empirical evidence derived from the regression model. For example, if a specific control factor is found to have a negligible or non-significant impact on the response variable, it may be considered for exclusion in future models, thus simplifying the model without sacrificing explanatory power. On the other hand, identifying key predictors that significantly influence the dependent variable can guide strategic interventions and optimizations in practical applications (Kulkarni et al., 2015). Overall, regression analysis serves as a fundamental tool in uncovering the complex interplay between different variables, thereby enhancing our understanding of the factors driving the observed outcomes and facilitating data-driven decision-making processes. Its applications extend across various fields, including economics, engineering, biological sciences, and social sciences, where understanding the relationships between variables is crucial for advancement and innovation. By leveraging regression analysis, researchers and practitioners can develop robust models that not only explain past data but also predict future trends, leading to more effective planning and resource allocation (Agarwal et al., 2021 Umali et al., 2020).

$$R = 3.28 - 0.0038TON - 0.0875TOFF - 0.0476WF - 0.000001WT - 0.0497UF - 0.0431LF \quad (3.2)$$

3.6 OPTIMIZATION TOOL USED

Optimization, as a strategic and systematic approach, plays a crucial role in achieving optimal outcomes under specific constraints or conditions. This methodical process involves a detailed assessment and enhancement of existing procedures to streamline operations and improve performance efficiency. Through a meticulous analysis of numerous factors and variables, optimization aims to minimize resource allocation while maximizing benefits and outcomes (Shendge & Shinde, 2015). By addressing complexities with innovative solutions and adaptability, optimization strikes a delicate balance between input efforts and output results (Saklecha et al., 2022).

Furthermore, the essence of optimization lies in its continuous journey toward refining methodologies and strategies to ensure the pursuit of the most effective approaches. This journey toward excellence involves wisely harnessing available resources to capitalize on opportunities and propel toward attainable goals (Patel et al., 2020). Optimization, therefore, represents a holistic and structured approach that aims to create synergy between various elements for optimal results. It requires a keen focus on identifying inefficiencies, streamlining processes, and constantly seeking ways to enhance performance (Kulkarni & Kakandikar, 2015).

Optimization goes beyond mere efficiency improvements; it also entails a mindset shift toward innovation and adaptability. This strategic approach enables organizations to not only achieve their current objectives but also adapt to changing environments and demands seamlessly. Through optimization, businesses can navigate complexities and unforeseen challenges with agility and foresight, ensuring that they are well-positioned for sustained success (Kaveh et al., 2023).

In essence, optimization serves as a cornerstone for organizations striving for operational excellence and competitive advantage. It embodies a commitment to continuous improvement, a dedication to leveraging resources effectively, and a willingness to explore new possibilities. By embracing optimization as a guiding principle, businesses can drive innovation, enhance productivity, and unlock new avenues for growth and success (Kulkarni et al., 2018, 2021).

3.6.1 TLBO Algorithm

Teaching is an important process, and everyone tries to learn something from others to improve themselves. An algorithm called TLBO has been proposed by Rao et al. It simulates traditional teaching and learning phenomena in the classroom. This algorithm simulates two basic learning modes. Through the teacher (called the teacher stage), (i) learners gain knowledge and skills from the teacher's expertise and guidance. (ii) They interact with other learners (called the learner stage) to enhance their understanding and capabilities through collaboration and peer learning (Rao & Patel, 2013).

"TLBO is a population-based algorithm in which a group of students (that is, learners) is considered a population, and the different topics offered to learners are similar to the different design variables of an optimization problem" (Rao & Patel, 2013). In this context, the learners represent potential solutions, and their performance or results are analogous to the objective values of the optimization problem. The process involves iterative refinement, where the knowledge imparted by the teacher helps the learners improve, and interactions among learners facilitate mutual enhancement of solutions.

The learner's result is similar to the applicability value of the optimization problem. The best solution for the total population is considered a teacher, who then influences the next generation of learners by sharing superior knowledge and strategies (Rao & Patel, 2013). This cyclical process mimics the educational cycle in a classroom, where continuous improvement is achieved through teaching and peer learning. The operation of the TLBO algorithm will be explained in the next teacher and learner stages (Patekar, 2016). During the teacher stage, the algorithm identifies the best-performing solution, which acts as the teacher. This teacher then guides the other solutions to improve by moving toward better positions in the solution space. In the learner stage, pairs of learners interact and learn from each other, potentially exchanging positions or creating new solutions through their interaction. This dual mechanism ensures a robust search process, balancing exploration and exploitation to find optimal or near-optimal solutions effectively. Overall, TLBO exemplifies how natural learning processes can inspire sophisticated optimization techniques, providing valuable tools for solving complex problems across various domains (Kakadikar et al., 2020b). Its success demonstrates the potential of educational metaphors in enhancing algorithmic efficiency and effectiveness.

3.6.2 Teacher Phase

"This stage of the algorithm simulates the learning of students (i.e. learners) through teachers" (Neve et al., 2020). At this stage, teachers strive to convey knowledge to learners and improve the average class performance (Kulkarni & Kakandikar, 2023b; Jawade et al., 2021). Suppose that m subjects (i.e., design variables) are provided for n learners (i.e., population size, $k = 1, 2, \ldots, n$). In a continuous educational learning cycle, i, M_j, and i are the average scores of learners in a specific topic j (j = 1, 2, . . ., M). The learner with the best total number is considered to be the algorithm teacher because the teacher is the most experienced and knowledgeable person in the subject. Let $X_{total\text{-}kbest,i}$ be the result of considering the best learners of all subjects identified as periodic teachers. Teachers do their best to improve the knowledge level of the entire class, but learners acquire knowledge according to the quality of education they provide and the quality of the class learners. Considering this fact, the difference between the teacher's grades and the average learners' grades for each subject is expressed as follows.

$$Difference\_Mean_{j,i} = r_i\left(X_{j,kbest,i} - T_F M_{j,i}\right) \tag{3.3}$$

where $X_{j,kbest,i}$ are the results of the teacher of topic j (i.e., the best learner). TF is the teaching factor to determine the average change, and r_i is a random number in the range of [0, 1]. The value of TF is 1 or 2. The value of TF is randomly determined with equal probability, as shown in the following equation (Kakandikar et al., 2022).

$$TF = round\left[1 + rand(0,1)\{2-1\}\right], \tag{3.4}$$

where *rand* is the random number in the range [0, 1]. TF is not a parameter of the TLBO algorithm in Eq. (3.4). The value of TF is not given as an input to the algorithm, and its value is randomly decided by the algorithm using Eq. (3.4). Based on the Difference_Mean$_{j,i}$, the existing solution is updated in the teacher phase according to the following expression (Pachpore & Jadhav, 2023):

$$X'_{j,k,i} = X_{j,k,i} + Difference\_Mean_{j,i} \tag{3.5}$$

where $X'_{j,k,i}$ are the updated values of $X_{j,k,i}$ in Eq. (3.5). If you get a better function value, accept $X'_{j,k,i}$. All function values accepted at the end of the teacher stage will be retained, and these values will be entered into the learner stage. Note that the values of r_i and TF will affect the performance of the TLBO algorithm. r_i is a random number in the range of [0,1], and TF is the teaching factor. However, the r_i and TF values are randomly generated by the algorithm and are not provided as inputs. The input parameters differ across algorithms—for example, crossover and mutation probabilities in the Genetic Algorithm, inertia weight and cognitive/social coefficients in PSO, and colony size and limit values in ABC. Therefore, there is no need to adjust RI and TF in the TLBO algorithm (different from GA crossover and mutation probability, PSO inertia weight, cognitive and social parameters, ABC colony size, and restriction adjustments). For TLBO to work properly, only general control parameters such as population size and number of generations need to be adjusted. These general control parameters are necessary for the normal operation of all optimization algorithms based on the totality (Saklecha et al., 2022). Therefore, TLBO can be called an algorithm without algorithm-specific parameters (Ambhore et al., 2024; Joshi et al., 2017; Kakandikar et al., 2020a).

3.6.3 Learner Phase

In this phase of the algorithm, the emphasis is placed on the vital aspect of student learning, wherein meaningful interactions among students play a pivotal role in knowledge acquisition and mastery. Through dynamic student–student engagement, learners have the invaluable opportunity to exchange ideas, perspectives, and insights, thereby facilitating a collaborative learning environment that fosters cognitive growth and deep understanding. By actively participating in discussions and interactions with their peers, students not only consolidate their existing knowledge but also have the chance to explore new concepts and information. This interaction offers a platform where knowledge flows freely, enabling learners to benefit from the diverse expertise and experiences of their fellow students. As students engage with those possessing a more profound understanding of certain topics, they are exposed to new perspectives and insights, leading to enhanced learning outcomes and the acquisition of fresh knowledge. Through these interactions, students are encouraged to think critically, approach problems from various angles, and broaden their knowledge horizons, ultimately contributing to a rich and fulfilling educational experience. The learning phenomenon at this stage is as follows: this phase creates a dynamic and interactive learning landscape where students are not only receivers of knowledge but active contributors to their own learning journey (Kakandikar et al., 2022; Ambhore et al., 2024). The collaborative nature of student–student interactions allows for a symbiotic exchange of ideas that goes beyond mere memorization and regurgitation of facts. As students engage in dialogue and debate with their peers, they are challenged to defend their viewpoints, consider alternative perspectives, and refine their arguments through constructive feedback. This process of intellectual exchange cultivates a deeper level of understanding as students are exposed to differing opinions and diverse approaches to learning. Furthermore, the cultivation of a supportive and inclusive learning community nurtures a sense of belonging and encourages students to express their thoughts openly without fear of judgment. In this way, the learning environment becomes a safe space for intellectual exploration and growth, where students can test their ideas and hypotheses without reservation. The active engagement of students in the learning process fosters a sense of ownership and responsibility for their own educational development, empowering them to take charge of their learning and become self-directed learners. Overall, the transformative power of student–student interactions in this phase of the algorithm is evident in the profound impact it has on shaping students' academic growth and fostering a culture of lifelong learning (Joshi et al., 2017).

Randomly select two learners, P and Q, such that $X'_{\text{total}-P,I} \neq X'_{\text{total}-Q,I}$, where $X'_{\text{total}-P,i}$ and $X'_{\text{total}-Q,i}$ are the updated values of $X_{\text{total}-P,i}$ and $X_{\text{total}-Q,i}$, respectively, at the end of the teacher phase.

$$X''_{j,P,i} = X'_{j,P,i} + r_i(X'_{j,P,i} - X'_{j,Q,i}), \text{ if } X'_{\text{total}-P,i} > X'_{\text{total}-Q,i}, \tag{3.6}$$

$$X''_{j,P,i} = X'_{j,P,i} + r_i(X'_{j,Q,i} - X'_{j,P,i}), \text{ if } X'_{\text{total}-Q,i} > X'_{\text{total}-P,i}. \tag{3.7}$$

Equations (3.6) and (3.7) are for maximization problems; the reverse is true for minimization problems.

Accept $X''_{j,P,i}$ if it gives a better function value.

3.7 RESULTS AND DISCUSSION

In analyzing the SR S/N ratio response presented in Table 3.3, it is evident that the pulse-off time stands out as being at the highest level labeled "1." This observation underscores the significance of the pulse-off time in influencing the SR metric. Conversely, the upper flash impact on SR appears to be minimal compared to other factors. The order in which additional parameters impact SR is as follows: pulse-on time (TON), undershoot (LF), wire feed (WF), and wire tension (WT). For optimal SR results as depicted in Table 3.3, the ideal combination of process parameters is denoted by A2B4C5D4E4F5. The symbols A, B, C, D, E, and F represent TON, TOFF, WF, WT, UF, and LF, respectively. Each symbol corresponds to a specific process parameter, while the associated numerical values represent their respective levels. Consequently, aiming to minimize the SR metric necessitates configuring TON at level 2, TOFF at level 4, WF at level 5, WT at level 4, UF at level 4, and LF at level 5. The optimization problem at hand is addressed through the implementation of the teaching-based learning optimization algorithm. This algorithm effectively identifies the optimal parameter levels by simulating the teaching and learning process, where the best solution (the "teacher") guides the population toward improved performance. The teaching phase focuses on moving the mean of the population toward the best solution, while the learning phase involves interactions among learners to enhance their solutions. Subsequently, the obtained optimization result is meticulously compared against the original S/N ratio reading to gauge the effectiveness of the optimization procedures employed. This comparison involves assessing the improvement in the SR metric, which indicates the success of the TLBO algorithm in enhancing the process parameters. By evaluating the pre- and post-optimization SR values, researchers can determine the extent to which the optimization algorithm has reduced the variability and improved the signal quality. In conclusion, the comprehensive analysis of the SR S/N ratio, combined with the strategic implementation of the TLBO algorithm, provides a robust framework for optimizing process parameters. This approach not only identifies the most influential factors but also prescribes their optimal levels to achieve superior performance. Through this methodology, significant improvements in SR can be realized, leading to more efficient and effective process outcomes. This iterative optimization and evaluation cycle exemplifies the powerful synergy between statistical analysis and algorithmic optimization in achieving high-quality results in complex engineering processes. Table 3.4 shows the results for the optimized parameters with TLBO algorithm.

3.8 CONCLUSION

The primary objective of the ongoing research project revolves around determining the optimum combination of single-response optimization process parameters that

TABLE 3.4
Result Table

Parameters	Results from S/N Ratio	Optimized Results for TLBO
TON	5	8
TOFF	3	7
WF	8	9
WT	900	900
UF	7	10
LF	7	9
SR	1.488	1.3229

are well-suited for processing high-carbon and high-chromium steels. This meticulous analysis is conducted employing the Taguchi method alongside the sophisticated TLBO optimization algorithm. In essence, the crux of this research endeavor lies in the precise customization of process parameters to cater specifically to the characteristics of these steels. Throughout the experimental design phase, the key to this investigative journey has been the systematic guidance provided by the L25OA, which has been instrumental in orchestrating the comprehensive exploration of myriad parameter configurations.

Initially, the analysis heavily leaned on the S/N ratio to meticulously identify the most effective settings for the process parameters that ultimately lead to the minimization of the single response. Notably, through empirical evidence, it was unequivocally established that TOFF emerged as the pivotal influencing factor that significantly contributed to SR outcomes. This finding highlighted the importance of optimizing the pulse-off time to achieve superior performance metrics. As the study transitioned into the subsequent phase, the incorporation of the SSA method was pivotal in addressing the optimization challenges that had been pinpointed earlier, thereby driving exceptional results to fruition. The SSA method, known for its robustness in handling complex optimization problems, was applied to refine the process parameters further and enhance the overall performance. Upon a comparative analysis between the SR values derived from the S/N ratio and those obtained through the SSA application, there emerged a conspicuous improvement in performance benchmarks. Impressively, the SR value extrapolated from the S/N ratio analysis stood at 1.488, and through the implementation of the SSA method, a remarkable enhancement ensued with a diminished SR value of 1.3229. This substantial 11% reduction in the SR metric precisely underscores the prowess and efficacy of the SSA approach in the overarching goal of reducing SR values, thereby fortifying its indispensable role within the framework of the optimization process.

The refinement of these analytical methods further underscores the meticulous attention to detail adopted throughout the investigative journey, solidifying the integrity and reliability of the research outcomes. By systematically applying and comparing different optimization techniques, the research not only identifies the most effective parameters but also demonstrates the value of combining multiple methodologies for enhanced results. This approach ensures that the final process parameters are finely tuned to achieve optimal performance, thereby advancing the state of knowledge in processing high-carbon and high-chromium steels. The success of this research highlights the critical role of advanced optimization algorithms and robust experimental designs in tackling complex industrial challenges, paving the way for future innovations in materials processing.

ACKNOWLEDGMENTS

This work is supported by all authors, and all authors have worked on the success of the completion of this chapter.

REFERENCES

Agarwal, V., Jawade, S., Atre, S., & Kulkarni, O. (2021). The role of mechanical testing in additive manufacturing. Material Science, Engineering and Applications, 1(2), 21–31. http://dx.doi.org/10.21595/msea.2021.22258

Ambhore, N., Gaikwad, M., Patil, A., Sharma, Y., & Manikjade, A. (2024). Predictive modeling and optimization of dry turning of hardened steel. International Journal on Interactive Design and Manufacturing, 18, 6281–6287. https://doi.org/10.1007/s12008-023-01615-y

Babbar, A., Rai, A., & Sharma, A. (2021). Latest trend in building construction: Three-dimensional printing. Journal of Physics: Conference Series, 1950(1), 012007.

Bhayana, M., Singh, J., Sharma, A., & Gupta, M. (2023). A review on optimized FDM 3D printed Wood/PLA bio composite material characteristics. Materials Today: Proceedings. https://doi.org/10.1016/j.matpr.2023.03.029

Bhoskar, T., Kulkarni, O. K., Kulkarni, N. K., Patekar, S. L., Kakandikar, G. M., & Nandedkar, V. M. (2015). Genetic algorithm and its applications to mechanical engineering: A review. Materials Today: Proceedings, 2(4), 2624–2630.

Bogar, M. N., Kulkarni, O., & Kakandikar, G. (2023). Micro forming studies of SS316L as biomedical application material. Journal of Engineering Science & Technology Review, 16(4).

Burande, C. G., Kulkarni, O. K., Jawade, S., & Kakandikar, G. M. (2021). Process parameters optimization by bat inspired algorithm of CNC turning on EN8 steel for prediction of surface roughness. Journal of Mechatronics and Artificial Intelligence in Engineering, 2(2), 73–85.

Gaikwad, M. U., Gaikwad, P., Krishnamoorthy, A., Jatti, V. S., Ambhore, N., Oza, A. D., . . . & Joshi, U. (2023). Fatigue analysis of electro discharge machined Nitinol 60. Innovation and Emerging Technologies, 10, 2340009. https://doi.org/10.1142/S2737599423400091

Gaikwad, M. U., & Jatti, V. S. (2020). Estimation of surface integrity parameters in electrical Discharge Machining (EDM) process-a review. In Techno-Societal 2018: Proceedings of the 2nd International Conference on Advanced Technologies for Societal Applications (Volume 2, pp. 601–614). Springer International Publishing. https://doi.org/10.1007/978-3-030-16962-6_61

Gaikwad, M. U., Krishnamoorthy, A., & Jatti, V. S. (2020). Investigation on effect of process parameter on surface integrity during electrical discharge machining of NiTi 60. Multidiscipline Modeling in Materials and Structures, 16(6), 1385–1394. https://doi.org/10.1108/MMMS-10-2019-0179

Jawade, S., Kulkarni, O. K., & Kakandikar, G. M. (2021). Parameter optimization of AISI 316 austenitic stainless steel for surface roughness by Grasshopper optimization algorithm. Journal of Mechanical Engineering, Automation and Control Systems, 2(2), 87–97.

Joshi, A. S., Kulkarni, O., Kakandikar, G. M., & Nandedkar, V. M. (2017). Cuckoo search optimization-a review. Materials Today: Proceedings, 4(8), 7262–7269.

Kakandikar, G. M., Dhage, R., Kulkarni, O., & Nandedkar, V. (2020a). Optimization of machining parameters of high-speed toolpath to achieve minimum cycle time for Ti-6Al-4V. In Advanced Manufacturing and Processing Technology (pp. 23–44). CRC Press.

Kakandikar, G. M., Kulkarni, O. K., Patekar, S., & Bhoskar, T. (2020b). Optimising fracture in automotive tail cap by firefly algorithm. International Journal of Swarm Intelligence, 5(1), 136–150.

Kakandikar, G. M., Nandedkar, V. M., & Kulkarni, O. K. (2022). Multiverse multi-objective optimization of thinning and wrinkling in automotive connector. In Computational Intelligence in Manufacturing (pp. 1–22). Woodhead Publishing.

Kaveh, A., Almasi, P., & Khodagholi, A. (2023). Optimum design of castellated beams using four recently developed meta-heuristic algorithms. Iranian Journal of Science and Technology, Transactions of Civil Engineering, 47(2), 713–725.

Khare, A., Kakandikar, G. M., & Kulkarni, O. K. (2022). An insight review on jellyfish optimization algorithm and its application in engineering. Research in Computer Science and Engineering, 9(1), 31–40.

Kulkarni, M. N., Patekar, M. S., Bhoskar, M. T., Kulkarni, M. O., Kakandikar, G. M., & Nandedkar, V. M. (2015). Particle swarm optimization applications to mechanical engineering-a review. Materials Today: Proceedings, 2(4–5), 2631–2639.

Kulkarni, O., & Kakandikar, G. M. (2015). Spring back optimization in metal forming using cohort intelligence. International Engineering Research Journal (IERJ), 2, 2938–2942.

Kulkarni, O., & Kakandikar, G. M. (2022). Formability assessment with microstructural investigations for zirconium 702 thin foils: Bio-material applications. Advances in Materials and Processing Technologies, 8, 2367–2377.

Kulkarni, O., & Kakandikar, G. M. (2023a). The investigations on formability of tantalum RO5200 thin foils: Bio-material. IOP Conference Series: Materials Science and Engineering, 1284(1), 012029. IOP Publishing.

Kulkarni, O., & Kakandikar, G. M. (2023b). Novel product design of tool for investigating formability with microstructural study of bio-material titanium grade-II thin foils. International Journal on Interactive Design and Manufacturing, 17(5), 2765–2775.

Kulkarni, O., & Kakandikar, G. M., & Nandedkar, V. M. (2021). Application of Salp Swarm algorithm to solve constrained optimization problems with dynamic penalty approach in real-life problems. In Metaheuristic Algorithms in Industry 4.0 (pp. 73–82). CRC Press.

Kulkarni, O., Kulkarni, N., Kulkarni, A. J., & Kakandikar, G. (2018). Constrained cohort intelligence using static and dynamic penalty function approach for mechanical components design. International Journal of Parallel, Emergent and Distributed Systems, 33(6), 570–588.

Kulkarni, O., & Kulkarni, S. (2018). Process parameter optimization in WEDM by Grey Wolf Optimizer. Materials Today: Proceedings, 5, 4402–4412.

Kulkarni, S., & Rodge, M. (2014). Process parameter optimisation in WEDM of HCHCR steel using Taguchi method and utility concept. International Journal of Mechanical Engineering and Technology (IJMET), 5(1), 57–67.

Kumar, A., & Singh, D. K. (n.d.). Strategic optimization and investigation effect of process parameters on performance of Wire Electric Discharge Machine (WEDM). International Journal of Engineering Science and Technology (IJEST). https://doi.org/10.1016/j.matpr.2023.06.113

Kumar, V., Babbar, A., Sharma, A., Singh, R., Pathri, B. P., & Prakash, C. (2024, February). Electric discharge machining: A comprehensive review for difficult to cut materials. In AIP Conference Proceedings (Vol. 2986, No. 1). AIP Publishing.

Mhatugade, S. P., Kakandikar, G. M., Kulkarni, O. K., & Nandedkar, V. M. (2019). Development of a multi-objective Salp Swarm algorithm for benchmark functions and real-world problems. In K. Kumar, & J. P. Davim (Eds.), Optimization for Engineering Problems (pp. 89–113). Wiley. https://doi.org/10.1002/9781119644552.ch5

Nene, A., & Kulkarni, O. K. (2022). Optimization of thermal efficiency of Scheffler solar concentrator receiver using slime mold algorithm. In Computational Intelligence in Manufacturing (pp. 71–86). Woodhead Publishing.

Neve, A. G., Kakandikar, G. M., & Kulkarni, O. (2017). Application of grasshopper optimization algorithm for constrained and unconstrained test functions. International Journal of Swarm Intelligence and Evolutionary Computation, 6(3). https://doi.org/10.4172/2090-4908.1000165

Neve, A. G., Kakandikar, G. M., Kulkarni, O., & Nandedkar, V. M. (2020). Optimization of railway bogie snubber spring with grasshopper algorithm. In Data Engineering and Communication Technology: Proceedings of 3rd ICDECT-2K19 (pp. 941–952). Springer Singapore.

Pachpore, S. S., & Jadhav, P. V. (2023). Endodontic treatment parameter optimization using central composite rotatable design. Research on Engineering Structures & Materials, 10(2), 679–689.

Patekar, L. (2016). Ant colony optimization and its applications in mechanical engineering review. Industrial Engineering Journal, 9(2), 41–45.

Patel, G., Kakandikar, G. M., & Kulkarni, O. (2022). Experimental and numerical investigations on forming limit curves in micro forming. Advances in Materials and Processing Technologies, 8(1), 33–44.

Patel, S., Kakandikar, G. M., & Kulkarni, O. (2020). Applicability and efficiency of socio-cultural inspired algorithms in optimizing mechanical systems-a critical review. Review of Computer Engineering Studies, 7(2).

Patnaik, A., & Mahapatra, S. S. (2006). Parametric optimization of Wire Electrical Discharge Machining (WEDM) process using Taguchi method. Journal of the Brazilian Society of Mechanical Sciences and Engineering, 28(4), 422–429.

Rao, R. V., & Patel, V. (2013). An improved teaching-learning-based optimization algorithm for solving unconstrained optimization problems. Scientia Iranica, 20(3), 710–720. https://doi.org/10.1016/j.scient.2012.12.005

Rao, S., & Venkaiah, N. (2013). Review on wire-cut EDM process. International Journal of Advanced Trends in Computer Science and Engineering, 2(6), 12–17.

Reddy, B. C., Sivakiran, S., & Eswarareddy, C. (2012). Effect of process parameters on MRR in wire electrical discharge machining of En31 steel. International Journal of Engineering Research and Applications (IJERA), 2(6), 1221–1226.

Saklecha, J., Pachpore, S., & Kulkarni, O. (2022). Generation of isometric projections in MATLAB. In Techno-Societal 2016, International Conference on Advanced Technologies for Societal Applications (pp. 927–936). Cham: Springer International Publishing.

Saklecha, J., Pachpore, S., Kulkarni, O., & Kakandikar, G. (2023). Comparative analysis of parameters affecting micro-forming process. Engineering Review: Međunarodni časopis, 43(2).

Sharma, A., Babbar, A., Singh, K., Singh, A. K., Tripathi, N. M., & Desai, D. J. K. (2024a). Effect of Minimum Quantity Lubrication (MQL) method on machining characteristics for ductile substrates: A future direction. In Modern Hybrid Machining and Super Finishing Processes (pp. 35–59). CRC Press.

Sharma, A., Grover, V., Babbar, A., & Rani, R. (2020b). A trending nonconventional hybrid finishing/machining process. In Non-Conventional Hybrid Machining Processes (pp. 79–93). CRC Press.

Sharma, A., Jain, V., & Gupta, D. (2020a). A novel investigation study on float glass hole surface integrity & tool wear using chemical assisted rotary ultrasonic machining. Materials Today: Proceedings, 26, 632–637.

Sharma, A., Jain, V., & Gupta, D. (2024b). In-situ experimental investigation using chemical-assisted rotary ultrasonic drilling process on fragile material. International Journal on Interactive Design and Manufacturing, 1–10.

Sharma, A., Singh, A. K., & Kumar, S. (2024). Performance evaluation of machining efficiency using graphite powder-based EDM: For aerospace application. International Journal on Interactive Design and Manufacturing (IJIDeM), 18(10), 7391–7400.

Sharma, S., Mehta, S., & Chopra, N. (2015). Economic load dispatch using Grey Wolf Optimization. International Journal of Engineering Research and Applications, 5(4), 128–132.

Shendge, D. K., & Shinde, D. B. (2015). Castellated beam optimization by using finite element analysis: A review. International Journal of Engineering Science, 4, 12–14.

Tiwari, N., Kakandikar, G., & Kulkarni, O. (2022). Micro forming and its application: A critical review. Journal of Engineering Research and Sciences, 1(3), 126–132.

Tupe, K., Zanje, V., Shaikh, S., Shigwan, A., & Kulkarni, O. (2017). Brief study on process parameter optimization in turning process: A review. International Journal of Innovative and Emerging Research in Engineering, 4(1).

Tupe, K., Zanje, V., Shaikh, S., Shigwan, A., Kulkarni, O., & Kulkarni, S. (2017). Optimization of cutting parameters for MRR in turning process of EN-31 steel using Taguchi approach. International Research Journal of Engineering and Technology (IRJET), 4(6).

Umali, R., Cruz, J. C., Fadchar, N. A., Manuel, M. C., Kim, S., Mahmud, A. R., Lee, W., & Messele, A. C. (2020). Design, fabrication and performance evaluation of a hybrid linear generator reciprocating-air compressor using a magnetic coupled piston. In 2020 11th IEEE Control and System Graduate Research Colloquium (ICSGRC) (pp. 236–241). IEEE.

4 Peering into the Smart Manufacturing—Material Forming and Joining

Casting and Welding to Unbind the Potential of Traditional Manufacturing

Sreejith S. Nair, Mahendra U. Gaikwad, and Kirubanidhi Jebabalan

4.1 INTRODUCTION

Industries are the building blocks of the national economy and its wealth, and improved performance and efficiency lead to a stronger economy. Newly established manufacturing businesses are increasingly embracing intelligent and smart technologies, which have been demonstrated to increase system productivity by 17–20% through enhanced machine utilization and energy optimization via smart manufacturing systems (SMSs) (Tuptuk et al., 2018). To meet consumer demands, businesses compete by inventing, reacting swiftly to market movements, keeping costs down, and producing dependable products, as also read in Ranjan et al. (2024). This competition has led to the advancement of smart manufacturing, process digitization, and cyberphysical control from manufacturing to retail outlets (Wang & Shih, 2016; Azizi et al., 2018). SMSs rely on electronic devices and sensors that require little human intervention, with customized manufacturing instructions provided directly from end customers via the Internet (Kang et al., 2016; Perera et al., 2013; Phuyal et al., 2020; Lydon, 2016). This revolution is driven by interconnected machining systems that enhance production planning, scheduling, product customization, agile manufacturing, and fault detection and recovery (Arnold et al., 2016). Industry 4.0 brings together advanced technologies like the Internet of Things (IoT), cyberphysical systems (CPSs), digital and smart manufacturing, additive manufacturing (AM), 5G communication, robotics, big data, data analytics, system integration, simulations, and flexible manufacturing systems (Scheer, 2015; Gunal, 2019; Fraga-Lamas et al., 2018; Mehrpouya et al., 2019; Lee, 2015; Ahuett-Garza et al., 2018; Al-Fuqaha et al., 2015). These innovative systems use cloud computing to connect product

DOI: 10.1201/9781003502647-5

design, analytics, manufacturing processes, inventory and supply chain management, product customization, real-time machining units, product delivery, and end-user engagement, resulting in a seamless, intelligent production landscape. This connectivity enables on-demand manufacturing, efficient product customization, and a more efficient demand-and-supply chain. The introduction of smart technology is transforming the production scene dramatically. Traditional material forming and joining processes, like casting and welding, are being revolutionized by cutting-edge technologies such as the IoT, artificial intelligence (AI), and data analytics (Sharma et al., 2024). This chapter explores the integration of smart technologies in material forming and joining, examines the benefits of smart manufacturing in traditional processes, provides case studies and practical applications, and discusses the issues and future prospects in this quickly evolving field. These strategies, which make use of data analytics and cutting-edge advancements, are helping manufacturers enhance efficiency, accuracy, and quality in previously inconceivable ways. Robotics and automation are being integrated into material forming and joining processes like as casting and welding to boost efficiency and precision to new heights thanks to advances in smart manufacturing (Sharma et al., 2020a, 2020b, 2024a, 2024b; Babbar et al., 2021; Bhayana et al., 2023). Manufacturers can eliminate waste and optimize production schedules by making proactive decisions based on instantaneous analysis of enormous amounts of data. This degree of innovation lowers costs and enhances quality control, making manufacturing processes more competitive and sustainable in the rapidly evolving market of today (Tripathi et al., 2024). Embracing cutting-edge technologies is essential when companies are trying to stay ahead of the competition and generate industry success.

This chapter examines several possible future directions in addition to providing a quick overview of the extent and significance of smart manufacturing solutions. The main focus is on how various technologies, CPSs, networked devices, and modern manufacturing system methods impact smart manufacturing. In this chapter, SMSs are presented, and their importance in several areas is highlighted. Furthermore, the integration of smart technology into the material shaping and joining processes is examined in detail in this chapter. Improved precision, efficiency, and flexibility are just a few of the many benefits that smart manufacturing offers over traditional production methods. The chapter provides detailed case studies and real-world applications to demonstrate how these innovative technologies have been successfully applied in a range of industries. Along with future directions and improvements in this rapidly evolving field, the chapter also highlights fresh concepts and innovations that could further revolutionize material forming and joining.

4.2 SMART MANUFACTURING SYSTEM

The SMS modifies the manufacturing landscape by digitally transforming all stages of the process (Lydon, 2016). It incorporates cutting-edge sensors and intensive big data analytics to enable continuous interoperability, real-time control and surveillance, adaptive manufacturing, rapid market responsiveness, and increased efficiency. SMS may operate in two modes: semi-autonomous and fully autonomous. In semi-autonomous mode, a production engineer sets goals and parameters to steer

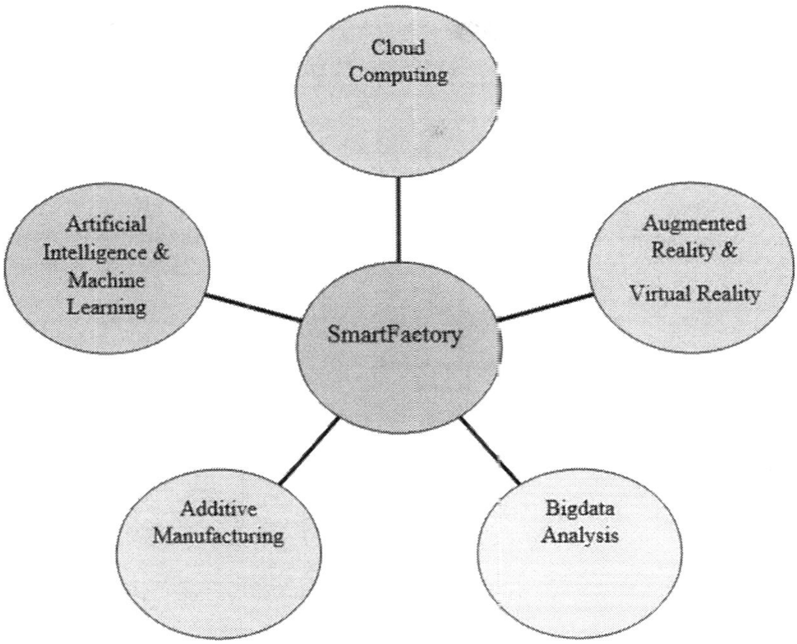

FIGURE 4.1 Components of a smart manufacturing system.

the production system toward peak performance. In fully autonomous mode, SMS takes instructions, determining optimal operating parameters and implementing them across all interconnected manufacturing units, resulting in a harmonious and intelligent production environment. Figure 4.1 illustrates visually the components of SMSs. Manufacturers' primary competitive concerns include cost-effectiveness, optimizing product manufacturing and delivery times, maintaining high product quality, and providing flexible product customization (Wang et al., 2014; Wang & Wang, 2018). Another key feature is the manufacturing system's ability to continuously enhance performance while utilizing real-time data and reacting to changing circumstances.

It improves manufacturing flexibility and customization, allowing for a faster reaction to market demands and lowering downtime. Furthermore, smart manufacturing increases resource utilization and sustainability, resulting in cost savings and environmental benefits. The Supply Chain Readiness Level and the Manufacturing Enterprise Solutions Association Manufacturing Transformation Strategy are two often-used tools for guiding this transformation. These strategies frequently focus on specific technologies or manufacturing execution systems rather than relying primarily on information and communication technology, which is necessary for evaluating and transitioning to intelligent systems. Table 4.1 shows the development plans for industrial infrastructure in globally developed countries.

TABLE 4.1
Industrial Development Strategies of Various Countries

S. No.	Country	Declared Strategy
1	Germany	Industry 4.0 (Scheer, 2015)
2	China	Made in China 2025 (Wang & Wang, 2016)
3	Japan	Society 5.0 (Li, 2018)
4	United States	Industrial Internet Consortium (Wang et al., 2016)

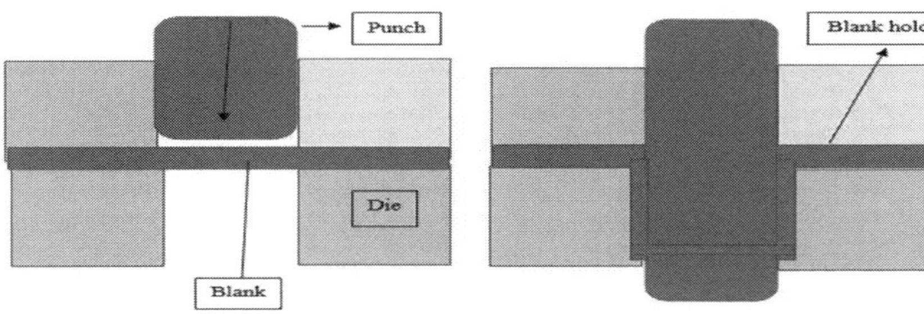

FIGURE 4.2 Deep drawing process.

4.2.1 Integration of Smart Technologies in Material Forming

Smart technologies provide real-time data on temperature, pressure, and other parameters, which enables exact control and optimization of the production process. Furthermore, smart technology aids predictive maintenance by detecting possible faults before they cause costly downtime. Adopting these innovations in material shaping and joining would catapult organizations to the forefront of their industries, increasing competitiveness and overall success. Moreover, integrating smart technology into material forming and joining processes enhances supply chain management. By leveraging data from sensors and monitoring systems, manufacturers can track the real-time status of raw materials and finished products. This increased insight enables better inventory management, less waste, and more efficient production planning.

4.2.2 JFE Intelligent Multistage Forming

JFE intelligent multistage forming (JIM-FORM) is a servo press-based deep drawing method for sheet materials. To allow lubricating oil to reflow between the die and the material's surface, the die and workpiece are intentionally separated during formation. The deep drawing mechanism is illustrated in Figure 4.2. The method prevents the lubricating oil layer from fracturing due to the strong contact pressure created by the workpiece sliding between the blank holder and the die. This is performed by slide position control and a servo-controlled die cushion, which moves the slide in pulses and changes the die cushion position accordingly. The pulse motion encourages the reinflow of lubricating oil while minimizing frictional forces. This method provides a higher margin and can manage a higher blank holding force (BHF) based on process conditions. However, deep sketching with pulse motion lengthens the process and lowers productivity. It is vital to monitor the lubrication condition during forming to maximize the relubrication impact while preserving production efficiency by using the proper motion settings based on the lubrication state.

Studies have explored enhancing the forming limit in deep drawing by utilizing low-frequency pulse vibration synchronized with the cushion and slide. This approach improves the forming limit by integrating low-frequency vibrations aligned with the motion of the cushion and slide. Each motion cycle of the vibration follows a two-stage pattern. During this period, the flange gap remains open, which reduces frictional and BHFs. To avoid wrinkles, it is crucial to keep the ram movement to a minimum. In step 2, the ram stops, and the cushion moves toward the ram, which helps reduce flange wrinkles. By utilizing this interlocking pulse motion, the forming height of the panel significantly increases, preventing breakage at the punch shoulder and reducing processing load. This approach maintains production efficiency by minimizing pulse width and increasing vibration frequency.

4.2.3 Challenges of JFE Intelligent Multistage Forming

Lubrication Management: An effective lubricant management system is one of the challenges in JFE's intelligent multistage forming. The procedure entails separating the workpiece and die during formation to permit lubricating oil to reflow, which can be difficult to effectively control. For successful formation, the lubrication process must be monitored and optimized to guarantee adequate lubrication and prevent film breakage.

Productivity versus Relubrication Effect: A significant obstacle in JFE intelligent multistage forming is striking a balance between the relubrication effect and productivity. Even while pulse motion reinflow of lubricating oil improves lubrication and lowers frictional force, it can cause longer process times and lower productivity. It is crucial to determine the ideal motion parameters to enhance the effects of relubrication while reducing the influence on production efficiency.

Monitoring System Integration: A challenge in JFE intelligent multistage forming is to integrate monitoring systems with die-embedded sensors, which are essential for detecting contact, friction, and shearing forces during the forming process. Examples of these sensors include ultrasonic, AE, and microdisplacement sensors.

4.3 INTELLIGENT WELDING SYSTEM

Intelligent welding employs advanced technology and machine learning (ML) to replicate, improve, or replace human expertise. By blending human skills with automated systems particularly in areas like computational analysis, precise

FIGURE 4.3 FANUC arc mate IoT-enabled robotic welding machine.

control, and sensing, welding processes become more efficient. This combination also makes it easier to manage, share, and apply human knowledge within these systems.

When human experience and knowledge are incorporated into the CPS through software and knowledge bases, the welding system's productivity, quality, and stability are consequently improved. An intelligent welding system (IWS) monitors and manages system, station, and system-of-systems (SoS) processes to accomplish its objectives. Table 4.2 displays the technological elements related to intelligent welding. The accuracy and depth of resource allocation are improved when numerous station-level IWSs are integrated via an industrial network to facilitate seamless data flow across greater areas. The industrial Internet and intelligent cloud platforms, which link different system-level IWSs, enable horizontal, vertical, and end-to-end integration that advances the industry at the SoS level.

IWSs frequently use theoretical, analytical, empirical, or numerical models to construct their robotic mechanisms and system models. The three types of these

TABLE 4.2
Concepts of Intelligent Welding

Concept	Definition	References
Robotic welding	Robotic welding refers to the whole management of the welding process through the use of automated, programmable equipment, which includes both weld execution and component manipulation.	Tarn et al. (2007)
Virtual mode welding	Virtual welding utilizes a robotic system to manage the torch and sensors for executing the weld and monitoring the weld pool surface. Simultaneously, the human operator uses a virtual welding torch designed to mimic a real one, enabling intuitive, unrestricted movement and adjustments in three-dimensional space.	Liu et al. (2014) and Liu (2016)
Welding 4.0 process	Welding 4.0 is a welding management system designed to enhance welding efficiency and minimize resource consumption.	www.ewm-group.com
Intelligent welding manufacturing	The term "intelligent welding manufacturing" refers to the use of artificial intelligence technology to replicate a welder's physical movements, cognitive abilities, and sensory perception.	Chen (2014) and Chen et al. (2014)

models are dynamic welding process models, characteristic parameters, and weld geometry descriptions. A picture of a robotic welding equipment with IoT capabilities made by FANUC Corporation Japan is shown in Figure 4.3. Models that can predict welding conditions and states are necessary for IWSs. Since they have an impact on weld quality, heat input, and material behaviour present significant issues for these models. Empirical models, commonly referred to as black box models, explain welding conditions using mathematical representations derived from physical observations. While they are effective in capturing system interactions, their application outside of the observed range may result in significant errors. Greater precision for modelling components and work cells is done through finite element models, which, when incorporated, offers many practical considerations.

4.3.1 Challenges of IWSs

Complexity of Integration: The integration of several technologies, including cloud computing, big data, AI, IoT, and intelligent manufacturing, is a key component of IWSs. It can be difficult to integrate these technologies effortlessly into a welding system because of compatibility problems and the requirement for specialized knowledge.

Data Security Concerns: The security of critical welding data becomes essential with the adoption of IWS technologies like cloud computing and IoT. To preserve the integrity of the welding processes, protecting data from cyber threats and unauthorized access is a major concern that must be handled.

Skill Upgradation: Transitioning to intelligent welding requires upskilling the workforce to operate and maintain these advanced systems effectively. Training employees to adapt to new technologies and processes can be a time-consuming and resource-intensive challenge for organizations.

Cost of Implementation: Implementing IWSs involves a significant initial investment in acquiring and setting up the required technologies. The cost of integrating these systems into existing welding processes can be a barrier for some organizations, especially smaller ones.

4.4 MECHANISM OF DIFFERENT AM TECHNOLOGIES USED IN CASTING

Casting is an effective method for making big metal parts. This has led to the development of rapid casting, a novel process that blends classical casting and AM. Through the use of rapid prototyping, different materials can be directly or indirectly employed to construct moulds, cores, or shells, which are then used in traditional casting methods to swiftly generate metal parts. Rapid casting overcomes the difficulties associated with mould fabrication in traditional casting by utilizing AM, leading to increased precision, quicker mould production cycles, and longer mould lifespans. This method is perfect for producing a variety of metal components in small batches while increasing manufacturing efficiency. It is widely used to make metal castings in the automotive and aerospace industries.

In the casting process, a number of AM techniques are commonly employed. Wax and resin-based moulds are made for investment casting using stereolithography (SLA) and selective laser sintering (SLS). Both the direct production of sand moulds and the printing of plastic and paper-based moulds for sand casting are common uses for fused deposition modelling (FDM) and laminated object manufacturing (LOM). By removing conventional processes, including core creation, moulding, pattern fabrication, and mould assembly, AM simplifies the process of creating casting moulds and greatly boosts production efficiency. A number of studies have contrasted the characteristics of traditional and fast casting. For example, Chen et al. combined AM and casting to address manufacturing and compositional homogeneity challenges in complicated multi-void ceramic components using porous alumina ceramic fibres for direct solidification and injection moulding (Chen et al., 2019). Past applications of 3D printing in casting have mainly focused on the rapid production of moulds, direct printing of metal products, and defect repair. An examination of the mechanisms behind AM application in casting shows that current AM techniques enable the swift creation of moulds, thereby enhancing casting productivity and reducing costs.

Figure 4.4 shows a photograph of the FDM machine used to create 3D components. One challenge with AM in casting is the direct deposition of powder using powder-feeding lasers on metal parts. Despite extensive research on forming modes and mechanisms, the material fusion mechanism, deformation behaviour, and forming laws of large metal components with various features processed through

FIGURE 4.4 Fused deposition modelling machine.

the integration of laser AM and casting are still not well understood. Additionally, there is limited knowledge about how thermal and mechanical coupling mechanisms impact the quality of the formed components. For multi-feature structures at the intersection of these two processes, the structural properties and design techniques need further investigation.

Even though studies on the technological and structural characteristics of casting and AM processes have been done, further research is required to provide structural design strategies that will allow these two processes to be integrated successfully. Moreover, there is a lack of consensus regarding the analysis and assessment techniques for forming the quality of components made using AM and casting integration. The current research on forming quality analysis and characterization methods for AM in casting parts is insufficient for effectively assessing or characterizing the forming quality of components produced through integrated processes. Table 4.3 lists the features and efficiency of the many AM processes that are applied to casting.

4.5 ROLE OF ROBOTICS AND AUTOMATION IN SMART MANUFACTURING

Robotics and automation play a crucial role in the advancement of smart manufacturing, revolutionizing how products are designed, produced, and delivered. In smart factories, robots and automated systems handle repetitive, complex, and hazardous tasks with unmatched precision and speed. This shift has allowed manufacturers to improve productivity while reducing errors and production costs. Collaborative robots or cobots work alongside human workers, enhancing safety and efficiency on the shop floor.

Figure 4.5 illustrates the role of robotics and automation in smart manufacturing, showcasing key elements such as collaborative robots, automated guided

TABLE 4.3
Comparison of Additive Manufacturing Process

S. No.	Additive Manufacturing Process	Material Used	Manufacturing Costs	Rate of Processing	Prototype Precision (mm)
1	Stereolithography (SLA)	Photosensitive resin	High	Fast	±0.1
2	Laminated object manufacturing (LOM)	Paper, plastic, and so on	Low	Relatively fast	±0.12
3	Selective laser sintering (SLS)	Coated sand, Paraffin, ceramic powder, and so on	High	Slow	±0.15
4	Fused deposition modelling (FDM)	Paraffin, nylon, ABS, PLA, and so on	Medium	Slow	±0.12

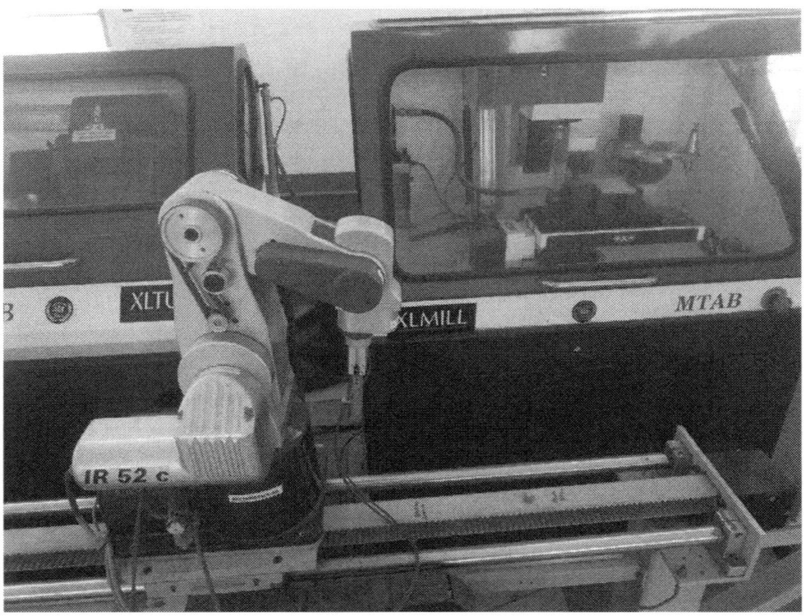

FIGURE 4.5 Robotics and automation in smart manufacturing.

vehicles (AGVs), AI-driven predictive maintenance, and automated production lines. Automated systems like computer numerical control machines, AGVs, and programmable logic controllers further streamline processes, improving overall output. AI enables automation by helping machines learn and optimize processes in real time, with AI-driven predictive maintenance systems monitoring equipment performance and predicting breakdowns to minimize costly downtime. The integration of robotics and automation has also enhanced product quality. Automated inspection systems with high-precision sensors ensure that every product meets strict standards. In industries like automotive, electronics, and pharmaceuticals, automation leads to faster, more accurate production cycles, meeting the high demand for customization and quality.

4.6 IMPACT OF INDUSTRY 4.0 ON THE SMART MANUFACTURING PROCESS

Industry 4.0, also known as the fourth industrial revolution is revolutionizing manufacturing by incorporating advanced technologies like the IoT, AI, ML, big data analytics, cloud computing, and robotics into production processes fostering "smart manufacturing" where interconnected systems, data-driven insights, and automation enhance efficiency, reduce costs, and improve product quality.

4.7 KEY TECHNOLOGIES OF INDUSTRY 4.0 IN SMART MANUFACTURING

Smart manufacturing processes are evolving as a result of the implementation of Industry 4.0 technology. The combined effects of these technologies result in a manufacturing ecosystem that is more flexible, integrated, and productive. This progression facilitates firms' ability to promptly address market demands, customize products, and sustain a competitive advantage in a swiftly evolving industrial topography. The applications of Industry 4.0 technologies in the manufacturing process are indicated in Table 4.4.

Table 4.5 highlights the implementation of Industry 4.0 technologies across various multinational corporations in their smart manufacturing processes. It highlights how leading companies are utilizing advanced technologies like IoT, AI, robotics, and automation to boost production efficiency, optimize resource utilization, and enhance product quality.

TABLE 4.4
Application of Industry 4.0 Technologies in Manufacturing Process

Technology	Application in Manufacturing
Internet of Things	Monitoring equipment health, tracking product movement, and optimizing supply chains
Big data analytics	Analysing production data for process optimization, predictive maintenance, and quality control
Cloud computing	Centralized data storage, analysis, and remote monitoring of production processes
Cyberphysical systems	Enabling smart factories where machines communicate with each other and make decentralized decisions

TABLE 4.5
Implementation of Industry 4.0 in Multinational Corporations (MNCs)

Company	Industry	Industry 4.0 Technology	Application in Smart Manufacturing
Siemens	Industrial equipment	IoT, digital twins, AI, automation	Energy consumption and product design
General Electric (GE)	Energy and manufacturing	AI, predictive maintenance, big data	Predictive maintenance
Bosch	Automotive and electronics	IoT, AI, robotics, automation	Automation of production lines
Caterpillar	Automotive	Robotics, AI, digital twins	Real-time defect detection

4.8 ADVANTAGES OF SMART MANUFACTURING IN TRADITIONAL PROCESSES

4.8.1 Improved Quality and Consistency

One of the main benefits of incorporating smart technologies into traditional manufacturing processes is the substantial enhancement in quality and consistency. Real-time monitoring and control ensure that process parameters stay within optimal ranges, thereby reducing the likelihood of defects. For example, in welding, maintaining consistent heat input and welding speed results in uniform welds with fewer defects. In casting, precise control over temperature and pouring speed minimizes variations in the cast parts, leading to higher-quality products. Digital twin technology makes it possible to run simulations that foresee possible problems, thus guaranteeing quality control all the way through the production cycle (Tyagi et al., 2023).

4.8.2 Enhanced Efficiency and Productivity

Smart manufacturing technologies streamline operations, resulting in enhanced efficiency and productivity. Automation reduces the time required for manual tasks, while real-time data analysis facilitates quick decision-making and process optimization. Automated welding systems enable continuous operation, significantly boosting production rates, while real-time monitoring and control in casting minimize downtime from manual inspections and adjustments, resulting in increased productivity.

4.8.3 Cost Reduction

The integration of smart technologies can result in substantial cost savings. Enhanced process efficiencies and reduced defect rates lead to lower material wastage and rework costs. Additionally, predictive maintenance enabled by IoT and AI can prevent unexpected equipment failures, thereby reducing maintenance costs and downtime. For instance, predictive maintenance in welding equipment can identify issues before they cause equipment failure, avoiding costly repairs and production stoppages. In casting, optimized process parameters minimize the need for rework, saving both material and labour costs. Smart manufacturing platforms utilize ML for predictive maintenance, optimizing processes based on empirical data, which can lead to significant cost savings (Ambrogio et al., 2024; Ambhore et al., 2024)

4.8.4 Enhanced Flexibility and Customization

Smart manufacturing allows for increased flexibility and customization in production. Advanced data analytics and AI enable manufacturers to respond fast to changes in product design and specifications. This capacity is especially useful in industries where customization and short manufacturing runs are commonplace. In welding, automated systems may be quickly reprogrammed to accommodate changing weld patterns and materials. In casting, data-driven insights allow for quick changes in mould designs and casting parameters to satisfy unique customer needs.

4.9 CASE STUDIES AND PRACTICAL APPLICATIONS

4.9.1 Case Study 1: Smart Welding in Automotive Manufacturing

In the automotive industry, welding is a critical process for assembling vehicle components. A major automaker incorporated IoT sensors and AI algorithms into its welding operations to improve quality and productivity. The IoT sensors tracked factors including current, voltage, and welding speed in real time. AI algorithms analysed the data and changed the welding parameters to ensure optimal circumstances. The integration of IoT sensors and AI algorithms in welding processes has revolutionized quality and efficiency in automotive manufacturing, as demonstrated by a case study in smart welding for car production. This approach, adopted by industry leaders like Toyota and Volkswagen, has reduced weld errors, increased manufacturing speed by 15%, and significantly boosted overall productivity. By enabling continuous monitoring of welding parameters such as current, voltage, and speed, IoT sensors, combined with AI-driven analytics, provide automotive manufacturers with a competitive edge through optimized operations, enhanced product quality, and data-driven decision-making. This real-time data gathering revealed useful insights into the welding process, allowing for early detection of anomalies or deviations from optimal circumstances. The firm was able to spot patterns and trends in this data using AI algorithms, which could then be utilized to dynamically modify welding conditions. For instance, if the AI detected a correlation between a specific combination of current and voltage and an increased likelihood of defects, it could automatically adjust these parameters to reduce the risk of defects.

The implementation of smart welding resulted in a 20% reduction in weld defects, a testament to the effectiveness of AI-driven optimization. By identifying and addressing potential issues proactively, the manufacturer was able to prevent defects before they occurred, reducing the need for costly rework or scrap. Additionally, the integration of IoT and AI led to a 15% increase in production speed. By optimizing welding parameters and minimizing downtime, the manufacturer was able to improve

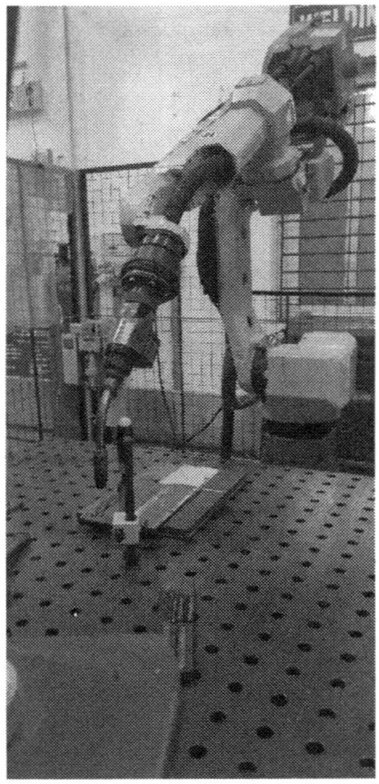

FIGURE 4.6 Robotic gas metal arc welding.

overall throughput and reduce manufacturing costs. Several research studies have highlighted the benefits of these technologies in enhancing operational efficiency, improving product quality, and reducing costs. Research demonstrated the potential of AI to revolutionize manufacturing processes through tasks such as predictive maintenance, quality control, and process optimization (Kusiak, 2023). Analysing the application of IoT in smart factories can lead to new ideas and advanced methods for enhancing quality control and optimizing part production processes (Soori et al., 2023). While the case study presents a compelling example of the benefits of smart welding, it is important to acknowledge that the implementation of such systems can be complex and challenging. Factors such as data quality, cybersecurity, and integration with existing infrastructure must be carefully considered. Figure 4.6 depicts the most common robotic gas metal arc welding used in the manufacturing of automotive components.

Valmet Car and Map Vision have unveiled a revolutionary flexible welding cell demonstration, marking a significant step forward in automobile manufacturing toward a self-adjusting production process. This one-of-a-kind closed-loop welding cell, created as part of Valmet Automotive's Intelligent Manufacturing programme, solves the challenges of a growing automotive sector that requires greater flexibility and capacity. This welding cell is driven by the Map Vision Smart Eye Cell, a cutting-edge system that combines several cameras, smart inspection software, and scalable hardware. The Map Vision Smart Eye Cell employs multi-camera technology and advanced inspection algorithms to precisely assess component placement and orientation prior to welding. The integration of AI and sensor systems improves welding precision and stability, as evidenced by case studies that show increased productivity and shorter cycle times (Kumar et al., 2024).

4.9.2 Case Study 2: IoT-Enabled Casting for Quality Management in Foundry

Aerospace components require high precision and quality, making casting a challenging process. The real-time data was analysed to verify that the molten metal remained within the correct temperature range, hence avoiding faults like porosity and shrinkage. This strategy resulted in a 30% reduction in errors and a 25% increase in casting efficiency, proving the value of smart solutions in high-precision industries. Figure 4.7 depicts an image of many high-precision, quality components made with sophisticated casting technology. In Korea, the die-casting industry is dominated by small and medium-sized firms (SMEs), which have limited financial resources and are hesitant to adopt smart factory principles. To solve this issue, the Korean government is actively encouraging SMEs to adopt smart industrial practices through a variety of support programmes. While these corporations have demonstrated a growing interest in implementing smart factories in accordance with government goals, the success of such facilities necessitates continuous monitoring. Although a monitoring system has been built, the collected data is not being utilized to its

FIGURE 4.7 High-quality cast component.

full potential. As a result, there is an urgent need to develop a system specifically designed to meet the unique requirements of the casting environment while ensuring efficient data processing.

Casting parameter data has been collected and is being analysed using modern data mining techniques to determine correlations between casting parameters and product quality. This project intends to provide die-casting enterprises that want to develop smart factories in the future with a methodical implementation architecture and protocols for properly managing casting parameters. The implementation of smart technologies has led to a 25% increase in casting efficiency, demonstrating the effectiveness of automation in high-precision industries (Stocker et al., 2024). Modernization of machinery, coupled with IoT integration, has been shown to enhance overall production quality while addressing new challenges in maintaining process stability (Patil Sachin et al., 2023). Furthermore, there is a desire for algorithms that can overcome the difficulty of multicollinearity among casting parameters, allowing for the development of unique products by selecting the best casting parameters. This comprehensive approach aims to move the die-casting industry into a new era of efficiency and innovation, laying the groundwork for long-term growth and competitiveness in the global market. IoT devices continuously monitor essential parameters like temperature and moisture content, both crucial for ensuring casting quality, and by collecting and analysing this data in real-time, foundries can promptly implement corrective actions leading to optimized energy consumption and improved overall operational efficiency (Srinivasan et al., 2022).

4.9.3 Case Study 3: Multistage Deep Drawing of Titanium

STÜKEN, a renowned manufacturer of precision metal components, possesses extensive expertise in cold forming challenging materials. After a detailed analysis of the requirements and adjustments to the manufacturing processes, an intelligent solution for the multistage deep drawing of titanium was developed. STÜKEN ensures high quality by using a precisely specified titanium alloy and accounting for the material's unique properties during processing. Despite the complex shaping process, the use of accurate high-performance tools enables the serial production of intricate titanium components. STÜKEN's methodology includes a detailed analysis of requirements and meticulous adjustments to manufacturing processes. The intelligent solution developed by STÜKEN allows for the efficient and reliable series production of complex titanium components. This capability opens up new possibilities for industries requiring high-performance titanium parts, such as aerospace, medical devices, and automotive sectors. In summary, STÜKEN's expertise in multistage deep drawing of titanium combines advanced material science, precise engineering, and stringent quality control to produce high-quality titanium components suitable for demanding applications. STÜKEN's innovative approach to multistage deep drawing of titanium components leverages advanced manufacturing techniques and precise material specifications. This methodology ensures high-quality production while addressing the unique challenges posed by titanium alloys. STÜKEN utilizes real-time monitoring and control systems to dynamically adjust process parameters, improving robustness against fluctuations in conditions like lubrication (Thiery et al., 2024).

4.10 CHALLENGES AND FUTURE DIRECTIONS

SMSs are intended to handle a variety of issues and complexities faced by current industries. However, their execution presents its own set of obstacles. These include security concerns, system integration challenges, a low return on investment (ROI) in new technology, and cost limits connected with developing new systems or upgrading old ones.

4.10.1 Cybersecurity Challenges in Smart Manufacturing

SMSs use an integrated network to share information between manufacturing units and end customers, generally over the Internet. Ensuring data security across the network is critical. This includes a worldwide unique identity and end-to-end data encryption to safeguard information at multiple stages throughout the system (Thoben et al., 2017). Each node must be protected from external threats and data exploitation. As a result, ensuring solid security throughout the process is an important issue when designing networked systems such as smart manufacturing. Innovative methods such as the Smart Reviving Authorization Model seek to improve security in chaotic systems by prohibiting unauthorized access and assuring steady operations (Huo & Wang, 2023). Research emphasizes the importance of adaptive defence strategies that can evolve with emerging threats, particularly in high-cost processes like semiconductor manufacturing (Dodge et al , 2023).

The integration of smart technologies in manufacturing brings advantages such as increased efficiency and productivity, but it also introduces new cybersecurity risks. To mitigate these threats, manufacturers must adopt proactive security strategies that protect sensitive data and systems while embracing innovation. A balance between technological advancement and robust security measures is crucial to safeguard operations and maintain trust.

4.10.2 System Integration

A significant challenge in implementing SMSs is integrating new technological equipment with existing systems (Phuyal et al., 2020). Compatibility issues often arise because older machinery operates on outdated communication protocols, whereas new devices may use different protocols. Effective machine-to-machine communication and system interconnectivity require advanced communication systems. Modern manufacturing systems often need IPv6 connectivity to support multiple devices simultaneously, which older systems may not support.

4.10.3 Interoperability

Interoperability refers to the dynamic ability of different systems, whether hardware or software-based, to understand and access each other's features independently, facilitating seamless data and information exchange (Hawaldar et al., 2018). In the context of Industry 4.0, interoperability is essential for ensuring that diverse systems can communicate and work together efficiently despite differences in their underlying technologies. It is categorized into three key levels:

Systematic Interoperability: This level focuses on adhering to the standards, procedures, models, and norms that are necessary for ensuring systems can be integrated seamlessly. It involves the establishment of universally accepted frameworks to allow consistent and reliable communication between systems.

Technical Interoperability: This involves managing the complexities of physical and digital communication between different systems. It ensures that systems, regardless of their hardware or software platforms, can exchange data effectively. It includes addressing network protocols, communication channels, and data formats to enable smooth interaction.

Semantic Interoperability: This aspect ensures that the shared information maintains its meaning across different systems. It focuses on making sure that data exchanged between systems is interpreted consistently, preserving its integrity and preventing errors due to misinterpretation of information.

Addressing these interoperability challenges is crucial for the effective implementation of SMSs. Ensuring interoperability across these four tiers allows for the seamless integration of various technologies, enabling smart factories to operate efficiently, meet modern production demands, and enhance security while maximizing performance. Without overcoming these challenges, the full potential of Industry 4.0 could remain untapped, as systems would be unable to work cohesively within the larger manufacturing ecosystem.

4.10.4 Safety in Human–Robot Collaboration

Human–robot collaboration, particularly with cobots or lightweight robots, opens up new possibilities in human–machine interface (HMI) by allowing for safe physical engagement between robots and humans in the workplace (Gualtieri et al., 2020; ISO, 2016) The International Federation of Robotics defines human–robot collaboration as robots' ability to coordinate with workers in industrial settings to accomplish specific tasks (Litzenberger et al., 2019). The primary concern in human–robot collaboration is the occupational health and safety of individuals who work closely together. It is vital to prevent harmful situations while adhering to the necessary occupational health and safety criteria. When adopting CPSs or industrial robotic systems, specific precautions should be taken to mitigate various workplace dangers such as mechanical, electrical, thermal, noise, vibration, radiation, and material/substance problems. Figure 4.8 highlights key elements of human–robot collaboration such as safe distancing, sensors, safety barriers, an emergency stop button, and a control system for monitoring.

4.10.5 Multilingualism

Systems need to be able to translate human language commands into machine language with ease to guide robots through desired tasks in the context of smart manufacturing (Uhlemann et al., 2020) These systems need to be able to take direct instructions from operators in text or speech language to successfully combine AI

FIGURE 4.8 Components of human–robot collaboration.

and advanced technologies and realize the goal of smart manufacturing. This function encourages a harmonic union of human intellect and machine efficiency while expediting operations and enabling human operators to engage with the industrial environment with ease.

4.11 ROI IN NEW TECHNOLOGY

Incorporating advanced technologies into modern industrial systems necessitates thorough financial analysis and ROI assessment (Patil Sachin et al., 2023). It is important to weigh the potential loss of production during the upgrading period against the larger investment required to adopt newer technologies. Whether or not to adopt newer technology depends largely on how long it takes to recoup the income from the old system and the ROI. The SMS has enormous commercial potential and is progressively gaining traction globally, per the conclusions of numerous research studies and publications from different organizations. Modern production systems, including reconfigurable ones, often lack vital features necessary for smart manufacturing, such as self-configuration, self-optimization, early detection, decision-making abilities, and predictive maintenance. The absence of these components restricts the systems' capacity to autonomously adjust, improve performance, and anticipate potential problems, thereby reducing their overall efficiency and effectiveness in contemporary manufacturing settings. Industrial automation systems' architecture is determined by their complexity, machine interaction and interrelation, subsystem roles, operational procedures, and interdependencies (Patil Sachin et al., 2023). Automation technology breakthroughs have increased the scope of data and information sharing between interconnected systems, leading to quick improvements in machine and process interoperability. This facilitates the management of adaptable resources and procedures as well as the appropriate handling of exceptions Phuyal et al., 2020)

Smart manufacturing, often referred to as intelligent manufacturing, is the next level of production that has been achieved via the application of automated and intelligent technology. By using cloud computing, these systems may quickly create customized products by accessing end-user data and information (Alcina, 2008; Harjunkoski et al., 2014). Fulfilling the demands of the modern market requires the implementation of reliable and automated solutions. Additionally, this technology has improved design alternatives, lowered production costs, and accelerated product delivery schedules. Strategic decision-making, psychological components, socioemotional wealth, and organizational commitment are some of the major factors that impact the complicated idea of ROI in the adoption of new technology. Companies looking to get maximum benefit out of their technological investments and guarantee long-term success need to understand these characteristics. Businesses can find themselves in a better position to make smart choices, lower risks, and increase the overall effect of their technology adoption programmes by properly evaluating these components.

Strategic decision-making is critical in determining the effectiveness of technological investments. Companies must first determine if they are "technology leaders" or "followers," as this status determines investment timing and risk tolerance. Krušinskas suggests that enterprises that take the lead in technology adoption may suffer higher risks but gain an early competitive advantage (Krušinskas et al., 2005). On the other side, organizations that follow and accept technology after it has been demonstrated may face fewer risks but miss out on prospective opportunities. This distinction highlights the importance of careful evaluation when determining the right time to invest in new technologies. In addition to timing, effective ROI justification requires a thorough understanding of both direct and indirect costs associated with technology adoption. Many companies focus solely on the direct costs, such as hardware, software, and implementation fees, without fully considering the indirect costs, such as training, process changes, and potential disruptions to existing workflows (Halpern et al., 2014). Moreover, there are often negative impacts that accompany new technology investments, such as unforeseen downtime, inefficiencies during the learning curve, and possible resistance from employees. Therefore, businesses must factor in these potential downsides when calculating the overall ROI of their technology investments to avoid underestimating the true cost and potential challenges they may face.

Another important aspect in technology adoption is the psychological qualities of decision-makers. Individuals in leadership positions' attitudes, goals, and attributes have a considerable impact on how and when new technology is implemented. According to Roberts et al., knowing these psychological aspects can provide useful insights into how organizations might improve their technology integration procedures (Roberts et al., 2021). Decision-makers who are driven by innovation and continual improvement are more likely to accept new technology, whereas those who are risk-averse or cautious about change may delay or even oppose adoption. Companies that recognize and address these psychological inclinations can customize their technology adoption tactics to coincide with their leaders' personal qualities, resulting in smoother transitions. Implementing a structured framework that categorizes these psychological elements can also help organizations foresee future bottlenecks and prepare for them ahead of time. Such frameworks can also help businesses choose the proper people to lead technology initiatives, focusing on persons with the psychological profiles that are best suited to managing technological change.

Socioemotional wealth, a term commonly associated with family-owned enterprises, complicates technology adoption decisions. In family businesses, socioemotional wealth refers to the non-financial characteristics that family members value, such as control, identity, and legacy. These considerations may occasionally prevent technology adoption, especially when there is a strong desire to uphold traditional customs and family control. According to Souder et al., family firms may be especially hesitant to adopt new technology if they fear these advances could damage their socioemotional prosperity (Souder et al., 2017). For example, the use of automation or digital technologies may be interpreted as a loss of personal control over specific operations, leading to reluctance to embrace these changes. When faced with the potential loss of market share or profitability, family businesses may be forced to reconsider their stance on technology adoption, striking a balance between preserving their socioemotional wealth and maintaining competitiveness. This highlights the delicate interplay between emotional and economic factors in the decision-making process, particularly in businesses where family values play a significant role.

The success of technology adoption is also significantly influenced by organizational commitment. Even the most promising technology investment may fall short of its potential if individuals inside the organization do not have a strong commitment to the project. The willingness of staff members at all levels to accept and incorporate new technologies into their regular work routines is a crucial component of successful technology adoption, according to Suebsin and Gerdsri (Suebsin et al., 2009). For this to occur, organizations need to cultivate a culture that values dedication to and excitement about technological progress. Effective communication is crucial to this process because it enables staff members to comprehend the long-term objectives, advantages, and purpose of the new technology. When employees are well-informed and feel included in the adoption process, they are more likely to take ownership of the technology and contribute to its successful implementation. Additionally, clear project scopes and well-defined roles are crucial for ensuring that all stakeholders understand their responsibilities and are aligned with the overall objectives of the technology initiative. By actively engaging stakeholders and encouraging open dialogue, companies can create a supportive environment where new technologies are embraced, rather than resisted.

In conclusion, a number of interconnected elements that affect how firms approach and implement technological innovations determine the ROI in the adoption of new technology. While leadership-related psychological aspects can help or hinder the adoption of new technology, strategic decision-making is crucial for identifying when to invest and evaluating potential dangers. Lastly, to guarantee that technology investments realize their full potential, organizational commitment and efficient communication are essential. Businesses may optimize the returns on their technology investments and set themselves up for long-term success in a competitive and increasingly digital environment by comprehending and tackling these aspects.

4.12 CONCLUSION

- The integration of smart technologies in material forming and joining processes such as casting and welding is transforming traditional manufacturing.
- The advantages of smart manufacturing, including improved quality, enhanced efficiency, cost reduction, and greater flexibility, are driving its adoption across various industries.
- Intelligent welding has enormous potential to revolutionize industrial processes, beginning with individual stations and seamlessly growing to include entire factory floors, thereby unlocking the full range of system integration benefits.
- The integration of AM with traditional casting techniques enhances productivity and reduces costs by enabling rapid and precise mould creation, particularly in industries requiring small-batch metal components.
- Industry 4.0 is transforming smart manufacturing through the integration of cutting-edge technologies that improve production flexibility, efficiency, and quality, allowing businesses to better satisfy consumer needs and stay ahead of the competition.
- SMSs face significant challenges, including security concerns, system integration difficulties, and the need for effective multilingual communication between humans and machines.
- Ensuring robust security, seamless interoperability, and safe human–robot collaboration will be essential for realizing the full potential of smart manufacturing.
- Incorporating advanced technologies into industrial systems requires a nuanced approach to financial analysis, balancing investment with potential production losses.
- As SMSs offer significant commercial benefits, addressing challenges such as ROI, psychological factors, and organizational commitment is crucial for successful implementation.
- Strategic decision-making and a comprehensive understanding of both direct and indirect costs will determine the effectiveness of these technological investments. Ultimately, companies that successfully navigate these complexities will strengthen their competitive advantage and secure long-term success in an increasingly dynamic digital landscape.
- Furthermore, the psychological attributes of decision-makers play a pivotal role in shaping technology adoption outcomes. Engaging employees, clarifying roles, and fostering a culture that values technological progress can enhance the overall impact of new technology investments.

- This work introduces an innovative technology that merges AM with casting processes to overcome current limitations. While challenges like integration complexity, data security, and the skills gap remain, the future of smart manufacturing holds great potential for producing large-scale metal components with advanced structures.
- Continued advancements in technology and a focus on training and upskilling will unlock the full potential of smart manufacturing, leading to more efficient, sustainable, and competitive manufacturing processes.

REFERENCES

Ahuett-Garza, H., & Kurfess, T. (2018). A brief discussion on the trends of habilitating technologies for Industry 4.0 and Smart manufacturing. Manufacturing Letters, 15, pp. 60–63. https://doi.org/10.1016/j.mfglet.2018.02.011.

Alcina, A. (2008). Translation technologies: Scope, tools and resources. Target. International Journal of Translation Studies, 20(1), pp. 79–102. https://doi.org/10.1075/target.20.1.05alc.

Al-Fuqaha, A., Guizani, M., Mohammadi, M., Aledhari, M., & Ayyash, M. (2015). Internet of things: A survey on enabling technologies, protocols, and applications. IEEE Communications Surveys & Tutorials, 17(4), pp. 2347–2376. https://doi.org/10.1109/COMST.2015.2444095.

Ambhore, N., Gaikwad, M., Patil, A., Sharma, Y., & Manikjade, A. (2024). Predictive modeling and optimization of dry turning of hardened steel. International Journal on Interactive Design and Manufacturing (IJIDeM), 18(8), pp. 6281–6287. https://doi.org/10.1007/s12008-023-01615-y.

Ambrogio, G., Filice, L., & Gagliardi, F. (2024). Smart manufacturing platform based on input-output empirical relationships for process monitoring. International Journal of Material Forming, 17(5), p. 45. https://doi.org/10.1007/s12289-024-01844-w.

Arnold, C., Kiel, D., & Voigt, K. I. (2016). How the industrial internet of things changes business models in different manufacturing industries. International Journal of Innovation Management, 20(8), p. 1640015. https://doi.org/10.1142/S1363919616400156.

Azizi, A., Yazdi, P. G., Humairi, A. A., Alsami, M., Rashdi, B. A., Al Zakwani, Z., & Al Sheikaili, S. (2018). Design and fabrication of intelligent material handling system in modern manufacturing with Industry 4.0 approaches. International Robotics & Automation Journal, 4(3), pp. 1–10. https://doi.org/10.15406/iratj.2018.04.00121.

Babbar, A., Rai, A., & Sharma, A. (2021). Latest trend in building construction: Three-dimensional printing. Journal of Physics: Conference Series, 1950(1), p. 012007.

Bhayana, M., Singh, J., Sharma, A., & Gupta, M. (2023). A review on optimized FDM 3D printed Wood/PLA bio composite material characteristics. Materials Today: Proceedings. https://doi.org/10.1016/j.matpr.2023.03.029.

Chen, C., Guo, B., Yu, Z., & Zhou, X. (2016). Internet plus in China. It Professional, 18(3), pp. 5–8. https://doi.org/10.1109/MITP.2016.47.

Chen, S. B. (2014). On intelligentized welding manufacturing. In International Conference on Robotic Welding, Intelligence and Automation (pp. 3–34). Cham: Springer International Publishing. https://doi.org/10.1007/978-3-319-18997-0_1.

Chen, S. B., & Lv, N. (2014). Research evolution on intelligentized technologies for arc welding process. Journal of Manufacturing Processes, 16(1), pp. 109–122. https://doi.org/10.1016/j.jmapro.2013.07.002.

Chen, Z., Li, J., Liu, C., Liu, Y., Zhu, J., & Lao, C. (2019). Preparation of high solid loading and low viscosity ceramic slurries for photopolymerization-based 3D printing. Ceramics International, 45(9), 11549–11557. https://doi.org/10.1016/j.ceramint.2019.03.024

Dodge, R., Pedrielli, G., & Jevtić, P. (2023). A testing based approach for security analysis of smart semiconductor systems. In 2023 Winter Simulation Conference (WSC) (pp. 2286–2297). IEEE. https://doi.org/10.1109/WSC60868.2023.10408730.

Fraga-Lamas, P., Fernandez-Carames, T. M., Blanco-Novoa, O., & Vilar-Montesinos, M. A. (2018). A review on industrial augmented reality systems for the Industry 4.0 shipyard. Ieee Access, 6, pp. 13358–13375. https://doi.org/10.1109/ACCESS.2018.2808326.

Gualtieri, L., Palomba, I., Wehrle, E. J., & Vidoni, R. (2020). The opportunities and challenges of SME manufacturing automation: Safety and ergonomics in human-robot collaboration. Industry 4.0 for SMEs: Challenges, Opportunities and Requirements, pp. 105–144. https://doi.org/10.1007/978-3-030-25425-4_4.

Gunal, M. M. (2019). Simulation for Industry 4.0. Past, Present, and Future, Springer, 3. https://doi.org/10.1007/978-3-030-04137-3.

Halpern, N. A., & Shaz, D. (2014). Return on investment: Just knowing the price is not enough!. Critical Care Medicine, 42(8), pp. 1952–1953. https://doi.org/10.1097/CCM.0000000000000422.

Harjunkoski, I., Maravelias, C. T., Bongers, P., Castro, P. M., Engell, S., Grossmann, I. E., Hooker, J., Méndez, C., Sand, G., & Wassick, J. (2014). Scope for industrial applications of production scheduling models and solution methods. Computers & Chemical Engineering, 62, pp. 161–193. https://doi.org/10.1016/j.compchemeng.2013.12.001.

Hawaldar, N., & Zhang, J. (2018). A comparative study of fabrication of sand-casting mold using additive manufacturing and conventional process. The International Journal of Advanced Manufacturing Technology, 97, pp. 1037–1045. https://doi.org/10.1007/s00170-018-2020-z.

Huo, X., & Wang, X. (2023). Internet of things for smart manufacturing based on Advanced Encryption Standard (AES) algorithm with chaotic system. Results in Engineering, 20, p. 101589. https://doi.org/10.1016/j.rineng.2023.101589.

ISO, S. (2016). Robots and Robotic Devices—Collaborative Robots (ISO-15066: 2016). International Organization for Standardization.

Kang, H. S., Lee, J. Y., Choi, S., Kim, H., Park, J. H., Son, J. Y., Kim, B. H., & Noh, S. D. (2016). Smart manufacturing: Past research, present findings, and future directions. International Journal of Precision Engineering and Manufacturing-Green Technology, 3, pp. 111–128. https://doi.org/10.1007/s40684-016-0015-5.

Krušinskas, R., & Vasiliauskaitė, A. (2005). Technology investment decisions to increase company value. Engineering Economics, 44(4), pp. 7–16.

Kumar, K. M., & Ojha, V. K. (2024). Optimization of GMAW processes in robotic welding: A case study from automotive manufacturing. International Research Journal on Advanced Engineering and Management (IRJAEM), 6(7), pp. 2288–2296. https://doi.org/10.47392/IRJAEM.2024.0332.

Kusiak, A. (2023). Predictive models in digital manufacturing: Research, applications, and future outlook. International Journal of Production Research, 61(17), pp. 6052–6062. https://doi.org/10.1080/00207543.2022.2122620.

Lee, E. A. (2015). The past, present and future of cyber-physical systems: A focus on models. Sensors, 15(3), pp. 4837–4869. https://doi.org/10.3390/s150304837.

Li, L. (2018). China's manufacturing locus in 2025: With a comparison of "Made-in-China 2025" and "Industry 4.0". Technological Forecasting and Social Change, 135, pp. 66–74. https://doi.org/10.1016/j.techfore.2017.05.028.

Litzenberger, G. (2019). IFR publishes collaborative industrial robot definition and estimates supply. International Federation of Robotics. IFR Secretariat Blog. Zugriff am, 24, p. 2020.

Liu, Y. (2016). Toward intelligent welding robots: Virtualized welding-based learning of human welder behaviors. Welding in the World, 60, pp. 719–729. https://doi.org/10.1007/s40194-016-0340-x.

Liu, Y., & Zhang, Y. (2014). Toward welding robot with human knowledge: A remotely-controlled approach. IEEE Transactions on Automation Science and Engineering, 12(2), pp. 769–774. https://doi.org/10.1109/TASE.2014.2359006.

Lydon, B. (2016). Industry 4.0: Intelligent and flexible production. International Society of Automation, pp. 12–17.

Mehrpouya, M., Dehghanghadikolaei, A., Fotovvati, B., Vosooghnia, A., Emamian, S. S., & Gisario, A. (2019). The potential of additive manufacturing in the smart factory industrial 4.0: A review. Applied Sciences, 9(18), p. 3865. https://doi.org/10.3390/app9183865.

Patil Sachin, S., Naik Girish, R., & Naik Poornima, G. (2023). Reinventing quality in foundry castings through gearbox housing optimization: A case study approach. Global Journal of Engineering and Technology Advances, 17(2), pp. 89–101. https://doi.org/10.30574/gjeta.2023.17.2.0231.

Perera, C., Zaslavsky, A., Christen, P., & Georgakopoulos, D. (2013). Context aware computing for the internet of things: A survey. IEEE Communications Surveys & Tutorials, 16(1), pp. 414–454. https://doi.org/10.1109/SURV.2013.042313.00197.

Phuyal, S., Bista, D., & Bista, R. (2020). Challenges, opportunities and future directions of smart manufacturing: A state of art review. Sustainable Futures, 2, p. 100023. https://doi.org/10.1016/j.sftr.2020.100023.

Ranjan, R., Singh, D., Pandey, V., Sharma, A., Sinha, A. S. K., & Tripathi, N. M. (2024). A comprehensive review on the machining process: Unconventional as an alternative to conventional machining. In Advances in Pre-and Post-Additive Manufacturing Processes (pp. 163–195). Boca Raton: CRC Press.

Roberts, R., Flin, R., Millar, D., & Corradi, L. (2021). Psychological factors influencing technology adoption: A case study from the oil and gas industry. Technovation, 102, p. 102219. https://doi.org/10.1016/j.technovation.2020.102219.

Scheer, A. W. (2015). Industry 4.0: From vision to implementation. Whitepaper [Online], 9. https://www.researchgate.net/publication/281447305_Whitepaper_-_Industry_40_From_vision_to_implementation.

Sharma, A., Babbar, A., Singh, K., Singh, A. K., Tripathi, N. M., & Desai, D. J. K. (2024a). Effect of Minimum Quantity Lubrication (MQL) method on machining characteristics for ductile substrates: A future direction. In Modern Hybrid Machining and Super Finishing Processes, (pp. 35–59). T aylor and Francis Group, CRC Press.

Sharma, A., Grover, V., Babbar, A., & Rani, R. (2020b). A trending nonconventional hybrid finishing/machining process. In Non-Conventional Hybrid Machining Processes (pp. 79–93). Taylor and Francis Group, CRC Press.

Sharma, A., Jain, V., & Gupta, D. (2020a). A novel investigation study on float glass hole surface integrity & tool wear using chemical assisted rotary ultrasonic machining. Materials Today: Proceedings, 26, 632–637.

Sharma, A., Jain, V., & Gupta, D. (2024). In-situ experimental investigation using chemical-assisted rotary ultrasonic drilling process on fragile material. International Journal on Interactive Design and Manufacturing (IJIDeM), pp. 1–10.

Sharma, A., Jain, V., & Gupta, D. (2024b). In-situ experimental investigation using chemical-assisted rotary ultrasonic drilling process on fragile material. International Journal on Interactive Design and Manufacturing, pp. 1–10.

Soori, M., Arezoo, B., & Dastres, R. (2023). Internet of things for smart factories in Industry 4.0, a review. Internet of Things and Cyber-Physical Systems, 3, pp. 192–204. https://doi.org/10.1016/j.iotcps.2023.04.006.

Souder, D., Zaheer, A., Sapienza, H., & Ranucci, R. (2017). How family influence, socio-emotional wealth, and competitive conditions shape new technology adoption. Strategic Management Journal, 38(9), pp. 1774–1790. https://doi.org/10.1002/smj.2614.

Srinivasan, N., Sricharan, A., & Shirish, K. (2022, October). Optimization of energy-related data: An IoT-enabled approach for metal casting industries. In AIP Conference Proceedings (Vol. 2653, No. 1, p. 040006). AIP Publishing LLC. https://doi.org/10.1063/5.0110843.

Stocker, T. F., Mehringer, J., Frechen, H., Sukowski, F., Schäfer, F., & Freier, D. (2024). Reduction of rejects by combining data from the casting process and automatic X-ray inspection. E-Journal of Nondestructive Testing, 29, p. 6. https://doi.org/10.58286/29884.

Suebsin, C., & Gerdsri, N. (2009, August). Key factors driving the success of technology adoption: Case examples of ERP adoption. In PICMET'09-2009 Portland International Conference on Management of Engineering & Technology (pp. 2638–2643). IEEE. https://doi.org10.1109/PICMET.2009.5261818.

Tarn, T. J., Chen, S. B., & Zhou, C. (2007). Robotic Welding, Intelligence and Automation (Vol. 362). Berlin: Springer. https://doi.org/10.1007/978-3-540-73374-4.

Thiery, S., El Abdine, M. Z., Heger, J., & Khalifa, N. B. (2024). Neural network-based estimation and compensation of friction for enhanced deep drawing process control. Materials Research Proceedings, 41. https://doi.org/10.21741/9781644903131-162.

Thoben, K. D., Wiesner, S., & Wuest, T. (2017). "Industrie 4.0" and smart manufacturing-a review of research issues and application examples. International Journal of Automation Technology, 11(1), pp. 4–16. https://doi.org/10.20965/ijat.2017.p0004.

Tripathi, N. M., & Sharma, A. (Eds.). (2024). Advances in Pre-and Post-Additive Manufacturing Processes: Innovations and Applications. CRC Press.

Tuptuk, N., & Hailes, S. (2018). Security of smart manufacturing systems. Journal of Manufacturing Systems, 47, pp. 93–106. https://doi.org/10.1016/j.jmsy.2018.04.007.

Tyagi, A. K., & Richa, R. (2023). Smart manufacturing using internet of things, artificial intelligence, and digital twin technology. In Global Perspectives on Robotics and Autonomous Systems: Development and Applications (pp. 184–205). IGI Global. https://doi.org/10.4018/978-1-6684-7791-5.ch008.

Uhlemann, J., Costa, R., & Charpentier, J. C. (2020). Product design and engineering—past, present, future trends in teaching, research and practices: Academic and industry points of view. Current Opinion in Chemical Engineering, 27, pp. 10–21. https://doi.org/10.1016/j.coche.2019.10.003.

Wang, L., & Shih, A. J. (2016). Challenges in smart manufacturing. Journal of Manufacturing Systems, 2(40), p. 1. https://doi.org/10.1016/j.jmsy.2016.05.005

Wang, L., & Wang, G. (2016). Big data in cyber-physical systems, digital manufacturing and Industry 4.0. International Journal of Engineering and Manufacturing (IJEM), 6(4), pp. 1–8. https://doi.org/10.5815/ijem.2016.04.01.

Wang, L., & Wang, X. V. (2018). Cloud-based Cyber-Physical Systems in Manufacturing (pp. 163–189). London: Springer. https://doi.org/10.1007/978-3-319-67693-7

Wang, Z., Chen, C., Guo, B., Yu, Z., & Zhou, X. (2016). Internet plus in China. It Professional, 18(3), pp. 5–8. https://doi.org/10.1109/MITP.2016.47

Wang, L., Wang, X. V., Gao, L., & Váncza, J. (2014). A cloud-based approach for WEEE remanufacturing. CIRP Annals, 63(1), 409–412. https://doi.org/10.1016/j.cirp.2014.03.114

5 Computational and Experimental Analysis of Erosion Wear in a Pipeline with Butterfly Valves

Laerson Nhantumbo, Gurmeet Singh, and Harjot Singh Gill

5.1 INTRODUCTION

Flow characteristics and particle erosion have garnered considerable interest in various industries such as energy, industrial, aerospace, oil and gas, chemical, automotive, and water distribution. The effective conveyance of fluids through fluid transport systems is one of the backbones of such sectors, where butterfly valves play an essential role in managing the fluid.

Butterfly valves, which are widely employed in hydraulic transportation, are frequently subject to damage from particle erosion. Equipment failure may result from severe erosion damage and fluid leaks. Hence, a crucial objective is to precisely forecast the corrosion properties and positions of butterfly valves to guarantee the safety of pipelines (Xu et al., n.d.a). Erosion–corrosion is widely recognized as a prevalent cause of pipeline failures.

Electrochemical corrosion has the potential to cause substantial deterioration of pipelines in oil and gas fields, leading to premature failure and the need for line replacement (Hu et al., 2011; Stack & Abdulrahman, 2012). Significant financial resources are allocated each year to mitigate the erosion–corrosion of pipelines stated by Wang et al. (2005). Therefore, computational fluid dynamics (CFD) modeling enables us to forecast and analyze possible erosion, hence facilitating the development of viable mitigation techniques to enhance the system's longevity.

An early study to investigate the erosion of butterfly valves in 1991 (ORNL/NRC/LTR-91/25, "Throttled Valve Cavitation and Erosion," 1991) discussed the causes, effects, and mitigation of cavitation and erosion in different types of valves used in power plants, including butterfly valves. Huang and Kim (1996) studied using the commercial code Fluent three-dimensional analysis of butterfly valve fluxes with partial opening, with a particular emphasis on the simulation results and the impact

DOI: 10.1201/9781003502647-6

of the mathematical model. Wojtkowiak and Oleśkowicz-Popiel (2006) examined the flow characteristics of butterfly valves by analyzing the valve design and conducting experimental and numerical analysis. The findings suggest an equation that enhances the flow characteristics. Song (2007) conducted a 3D numerical simulation using CFX to analyze the flow pattern, measure valve flow, and hydrodynamic torque coefficient at various opening angles. Song compared the results to real-world experimental data and discovered that the data obtained using CFX is consistent with the experimental data. In this research Del Toro et al. (2015) employed CFD to study and forecast the butterfly valve efficiency factor, considering various degrees of valve opening. The CFD results were then compared to experimental data. It was shown that valve angles between 70 and 90 degrees yielded reasonably accurate predictions, in contrast to smaller valve angles. Liu et al. (2017) concluded that, when velocity increases, the erosion rate of particle mass and size increases. The standard k-model, which is one of the RANS equation-based turbulence models included in ANSYS CFX, was employed to simulate the complex flow around a butterfly valve. This model is numerically stable and performs well in terms of prediction. It also boasts a well-established flow regime investigated by Duan et al. (2018). This research by Hee Lee and Ho Bae (n.d.) utilized CFX Software to model and analyze the flow properties surrounding the butterfly valve at varying disc angles; the mathematical model's validation is confirmed through comparison with experimental data. Xu et al. (n.d.b) conducted a study on erosion wear in a pipeline with a butterfly valve, providing some insights into the cause of erosion, the opening angle effect on the velocity of the flow, and the influence of particle magnitude on erosion. The study of Kyriakopoulos et al. (2022) employed both numerical and physical modeling techniques to verify the results obtained from a real-life experiment focused on the hydraulic performance of Howell–Bungers and butterfly valves in large dams; the findings indicate that the distance between the butterfly and Howell–Bungers valve should be a minimum of four times the diameter of the pipe. Li et al. (2023) used CFD-DEM to evaluate the erosion wear characteristics of a triple eccentric butterfly valve with graphite sealing surfaces, altering target materials, circumstances, particle sizes, and particle velocities, finding that the extent of deterioration reduces as the temperature of the surrounding environment increases. In another work, Zhang et al. (2023) offered an optimal design and operation for butterfly valves and provided an understanding of cavitation evolution in a valve. Perera et al. (2024) identified the valve as a whole and parts of the butterfly valve most sensitive to erosion. A few additional research papers concerning CFD and erosion wear that were not explicitly associated with butterfly valves were examined. Pandya et al. (2017) developed a reliable erosion prediction model by bringing together ANN with CFD. Zhang et al. (2022) predicted erosion wear in pipe interfaces through the use of numerical analysis and erosion modeling. The reviews shown here were conducted to provide a concise summary of the current state of CFD and erosion wear analysis.

Previous research papers have examined the impact of fluid flow, erosion, corrosion, and other factors on butterfly valves. However, these papers have not thoroughly addressed the intricate aspects of butterfly valves. Therefore, there is a need for further investigation. Specifically, there is a lack of research on different valve types, materials, and conditions, as well as various turbulence models. The effects of fluid flow on both the valve and the pipeline require more attention. Furthermore, there is potential for additional studies in specific areas, such as exploring wider particle sizes or different shapes, examining different types of liquids, investigating high-temperature applications, and assessing corrosion resistance. The study seeks to analyze fluid dynamics and erosion wear by comparing two simulations. Both simulations use the same data but differ in the magnitude of sand particles. There has been no previous study examining multiple butterfly valves in a single pipeline. Such a study can offer a fresh perspective and comprehensive analysis of flow characteristics and erosion, which can possibly disprove or offer knowledge that can be beneficial for other research connected to different fluid transport systems.

5.2 MODEL AND SIMULATION MODELING

5.2.1 Model Description

Figure 5.1 shows the schematic of the three butterfly valves within the pipeline with different degrees of rotation. The length of the pipe is 1440 mm, the diameter is 50 mm, and the thickness is 5 mm, and the butterfly valve has a diameter of 50 mm and a thickness of 6 mm. Each intake is located at a distance from the valve that is seven times the diameter. Similarly, the distance between each valve and the adjacent valve, as well as between the valve and the outlet, is at least seven times the diameter. The pipe was initially created using SolidWorks 2020 and subsequently imported into ANSYS 2020 R1 to generate the mesh domain and conduct the simulation.

5.2.2 Mesh and Simulation Modeling

Figure 5.2 shows how the fluid domain's mesh, which included the butterfly valves, was created using Space Claim. The inlet and outlet are identical for all models. Karimi et al. (2017) sought to improve the prediction of fine particle erosion by incorporating a variety of boundary layer types nonetheless, regarding these models the

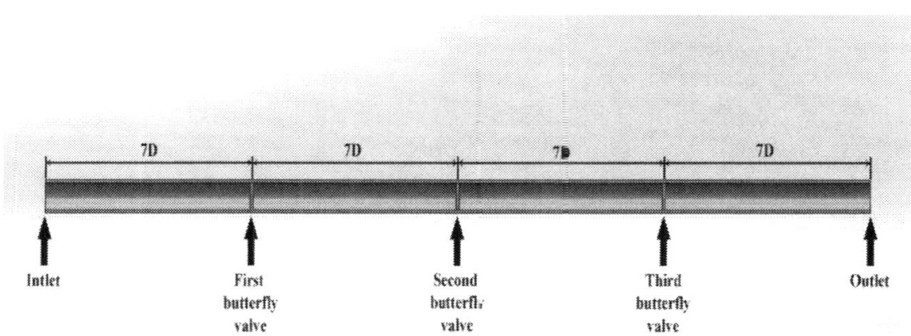

FIGURE 5.1 Representation of the base model

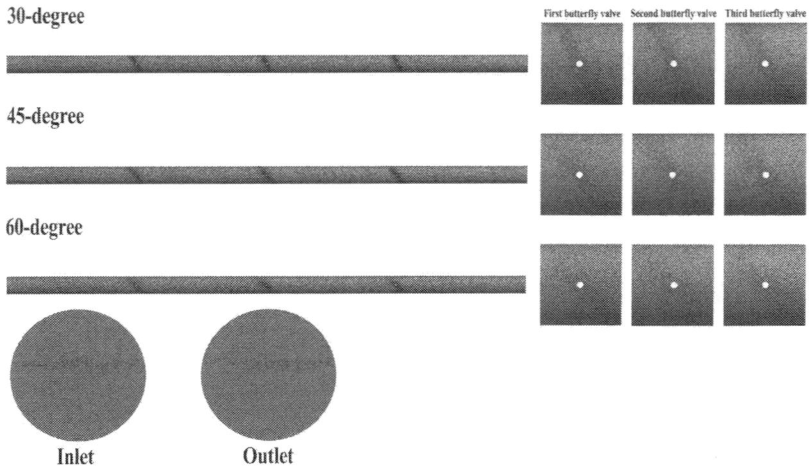

FIGURE 5.2 Mesh of the fluid domain.

division of the domain into smaller portions makes it so the quality of the mesh decreases and the erosion rate results are unreliable; therefore, mesh quality 0.05 m was chosen to generate the mesh along with an inflation of five layers. Finally, the respective nodes and elements for the 30, 45, and 60-degree models were 264139 and 737502, 258545 and 761921, 256588 and 779456.

An inlet velocity of 8.5 m/s is applied in the x-direction, with a pressure outlet set at 0 Pa. The working fluid, liquefied natural gas (LNG), has a density of 500 kg/m^3 and a viscosity of 0.00029 kg/(m·s). The solid material is 9% nickel steel with a density of 7890 kg/m^3. Sand particles are introduced from the surface with an initial velocity of 8 m/s in the x-direction, a density of 1600 kg/m^3, and a total mass flow rate of 1 kg/s.

Two simulations were performed to undertake this research: (a) and (b); the data utilized in both simulations are the same, with the only difference being the magnitude of the sand particles. For simulation (a) minimum particle diameter is 0.002 m, maximum particle diameter is 0.0045 m, and the mean particle diameter is 0.004 m, and for simulation (b) minimum particle diameter is 0.005 m, maximum particle diameter (*m*) is 0.0085 m, and the mean particle diameter is 0.007 m. The Rosin-Rammler distribution was selected to determine the particle size.

The models for CFD-discrete phase method (DPM) and CFD-DEM are commonly employed techniques for simulating interactions between fluids and particles (Elaswad et al., 2024). The DPM is particularly suitable for dilute particle fluxes and requires less computational resources (Fernandes et al., 2018). Therefore, CFD-DPM was chosen to carry out the simulation. One noteworthy investigation employing DPM was conducted by Mohebi et al. (2024).

The turbulence model employed is the viscous SST k-Omega model; k–ω was chosen according to Guan Song and Park (n.d.). The k–ω model provides higher accuracy for boundary level flows, while the k–ε model tends to overstate turbulence. The acceleration due to gravity is −9.81 m/s^2 in the *y*-direction. The coupled solution method is used to solve fundamental equations such as momentum and continuity. It is initialized in a hybrid manner under steady-state conditions.

5.3 MATHEMATICAL METHODS

The flow around the butterfly valve was assumed to be steady, incompressible, turbulent, and single-phase. In this study, ANSYS Fluent was used to perform the numerical calculations, and the governing equations were based on those provided in the ANSYS Fluent Theory Guide.

5.3.1 Governing Equations

The governing equations which were used for this research are:

Continuity equation,

$$\frac{\partial \rho}{\partial t} + \nabla \cdot (\rho \vec{v}) = S_m \tag{5.1}$$

where ρ is density, t is time, μ is the fluid vector, and S_m is the mass from the second phase dispersion added to the continuous phase.

Momentum equation,

$$\frac{\partial}{\partial t}(\rho \vec{v}) + \nabla \cdot (\rho \vec{v}\vec{v}) = -\nabla' p + \nabla \cdot \left(\overline{\overline{\tau}}\right) + \rho \vec{g} + \vec{F} \tag{5.2}$$

where $\overline{\overline{\tau}}$ is the stress tensor, $'p$ is the static pressure, $\rho \vec{g}$ is the body force gravitational, and $\vec{F}$ represents the external force of the body.

5.3.2 k–ω SST Model

The k–ω SST model is employed to incorporate turbulence effects and calculate the turbulent viscosity. The formulation of this model is as follows:

$$\frac{\partial}{\partial t}(\rho k) + \frac{\partial}{\partial x_i}(\rho k u_i) = \frac{\partial}{\partial x_j}\left(\Gamma_k \frac{\partial k}{\partial x_j}\right) + \tilde{G}_k - Y_k + S_k \tag{5.3}$$

and

$$\frac{\partial}{\partial t}(\rho \omega) + \frac{\partial}{\partial x_i}(\rho \omega u_i) = \frac{\partial}{\partial x_j}\left(\Gamma_\omega \frac{\partial \omega}{\partial x_j}\right) + G_\omega - Y_\omega + D_\omega + S_\omega \tag{5.4}$$

In these equations, $\tilde{G}_k$ is the kinetic energy of turbulence creation caused by mean gradients of velocity. G_ω denotes the creation of ω. Y_k and Y_ω represent the dissipation of k and ω due to turbulence. Γ_k and Γ_ω represents the efficacy of diffusivity for k and ω, respectively. D_ω is the term of cross-diffusion. S_k and S_ω are defined by users' terms of source.

5.3.3 Discrete Phase Model

This numerical model DPM employs the kω–*SST* RANS model to simulate turbulence of the flow. The DPM is then utilized in the fluid domain to examine the behavior of the dispersed phase while monitoring their motions in a Lagrangian frame. The continuous phase in a DPM model, however, is solved in an Eulerian frame; therefore, certain research may refer to DPM as an Eulerian–Lagrangian technique (Perera et al., 2024).

The force equilibrium equation on the particle is addressed by applying the local continuous phase conditions, as outlined by the succeeding equation:

$$\frac{dv_p}{dt} = F_D\left(v_f - v_p\right) + g\frac{\left(p_p - p_f\right)}{p_p} + F_x \tag{5.5}$$

Let v_p represent the velocity of the particle and v_f represents the velocity of the fluid. Similarly, let p_p represents the density of the particle and p_f represents the density of the fluid. g represents the gravitational acceleration. F_x is a term that takes into account any extra forces. Lastly, $F_D\left(v_f - v_p\right)$ represents the drag force per unit particle mass.

5.3.4 Finnie Erosion Model

Finnie (1960) and Finnie and Kabil (1965) suggested that erosive wear occurs when surfaces are cut by hitting particles. Although the Finnie model is old, it remains one of the most commonly utilized erosion models for ductile materials (Hankeln & Riedelbauch, 2019). Given the widespread usage of the Finnie model for ductile materials, it was selected as the model of choice for this work. The erosion rate is defined as:

$$E = kV_P^n f(\gamma) \tag{5.6}$$

The dimensionless erosion mass is denoted as E, the model constant is represented by k, the particle impact velocity is denoted as V_P , and the impact angle γ is described by a dimensionless function $f(\gamma)$. The exponent value for metals typically falls within the range of 2.3 to 2.5 radians.

Finnie's erosive wear model establishes a relationship between the pace at which particles contact the surface and their kinetic energy, utilizing the following functions:

$$f(\gamma) = \begin{cases} \frac{1}{3}\cos^2\gamma & \text{for } \gamma > 18.5^\circ \\ \sin(2\gamma) - 3\sin^2\gamma & \text{for } \gamma \leq 18.5^\circ \end{cases} \tag{5.7}$$

5.4 RESULTS AND DISCUSSIONS

5.4.1 Velocity Magnitude

According to Figure 5.3, for all degrees of rotation, the velocity magnitude increases at the margins or regions of the butterfly valve in both simulations (a) and (b).

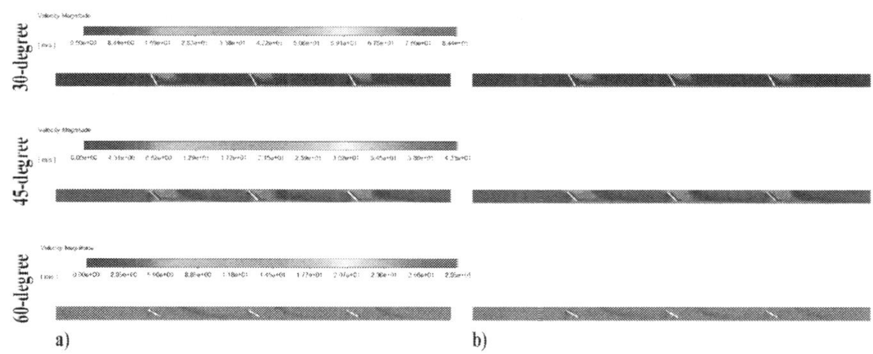

FIGURE 5.3 Velocity magnitude comparison.

A narrower opening will result in higher velocity magnitude on the valve area, which will then create a higher pressure drop (more energy loss), which is in line with the results in Luaibi and Abdulwahid (2022); they say that, with an increase in velocity, the pressure drop increases, which is not preferable as it can reduce efficiency, increase wear and tear, and potentially increase temperature, which can degrade the material faster.

From Table 5.1, particle size may not have a significant effect on the flow since the amount of velocity magnitude in both simulations at each angle of rotation is essentially identical around all butterfly valves. Nevertheless, simulation (a) consistently displays a higher velocity magnitude, and simulation (b) occasionally experiences a greater degree of fluctuation in the velocities of each valve, indicating that the fluid can be influenced by particle size in a variety of ways implying that the velocity magnitude is enhanced as a consequence of the larger particle size in simulation (a). These findings align with the observations presented in Gella et al. (2017).

5.4.2 Static Pressure

According to Figure 5.4, because the valve is confining and slowing down the fluid, there is a noticeable increase in pressure as the fluid is restricted by the valve and an increase in pressure drop after each valve in both simulations. The narrower

TABLE 5.1
Velocity Magnitude Comparison Maximum Value (m/s) Data

	30-degree		45-degree		60-degree	
Simulation	(a)	(b)	(a)	(b)	(a)	(b)
First butterfly valve	8.44e+01	8.43e+01	4.23e+01	4.30e+01	2.74e+01	2.60e+01
Second butterfly valve	8.44e+01	8.01e+01	4.23e+01	4.31e+01	2.93e+01	2.95e+01
Third butterfly valve	8.43e+01	8.00e+01	4.22e+01	4.30e+01	2.92e+01	2.94e+01

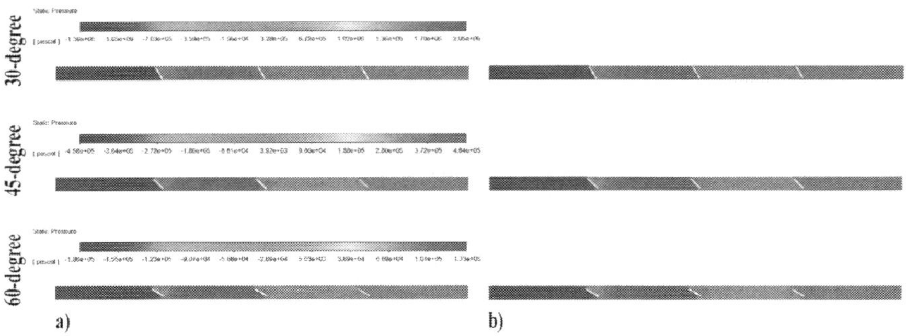

FIGURE 5.4 Static pressure comparison.

TABLE 5.2
Static Pressure Maximum Value (Pa) Data Comparison

	30-degree		45-degree		60-degree	
Simulation	**(a)**	**(b)**	**(a)**	**(b)**	**(a)**	**(b)**
From the inlet to the first butterfly valve	1.90e+06	2.05e+06	4.36e+05	4.64e+05	1.24e+05	1.33e+05
From the first butterfly valve to the second	1.25e+06	1.38e+06	3.00e+05	3.21e+05	9.29e+04	9.69e+05
From the second butterfly valve to the third	6.03e+05	7.27e+05	1.44e+05	1.76e+05	5.75e+04	5.98e+04

the opening angle, the larger the static pressure, which may accelerate mechanical stresses and potential damage due to cracking and fatigue and valve wear and tear. Just like in a separate form of fluid transport system, the retention shelter was damaged due to hydraulic impact or water pressure generated by a fluid (Janoszek & Rotkegel, 123 C.E.). Water hammer on the valves can damage the pipe or equipment due to the force exerted by the pressure (Sotoodeh, 2023).

The first butterfly valve is met with higher pressure compared to the others, which means potentially higher stress, meaning more damage to the first butterfly valve.

According to Table 5.2, simulation (b) exhibits higher static pressure compared to (a) across all degrees of rotation. This disparity in static pressure might result in more turbulence and flow disruption caused by bigger particles (Leskovec et al., 2020), ultimately leading to higher static pressure levels, not to mention potential obstructions created by the larger particles, which may have a huge effect on the increase of static pressure.

5.4.3 Erosion Wear Analysis

Figure 5.5 shows the erosion wear pattern in the pipeline with the butterfly valves, using the Finnie erosion model, where most of the butterfly valves have been affected

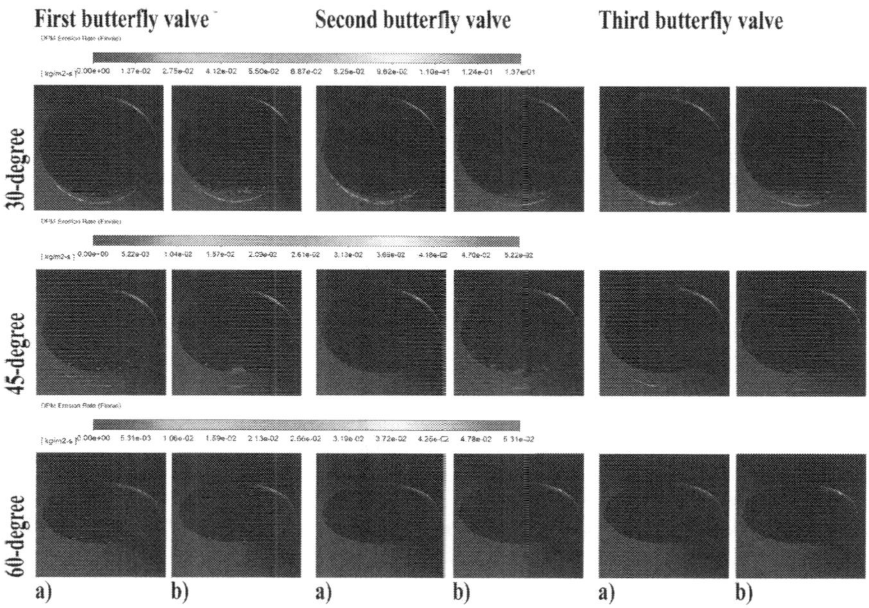

FIGURE 5.5 Comparison of the erosion rate on the pipe and butterfly valves.

primarily along their edges—an observation that aligns with Nystad et al. (2010), who noted that hydraulic oil contaminated with solid particles frequently erodes the corners of valve components—and the effects of gravity are more pronounced in simulation (b), which shows extensive erosion on the bottom of the valve and pipeline compared to simulation (a).

Figure 5.6 shows the erosion rate plot created using the Finnie model; the three peaks represent the rate of erosion on the walls around the first, second, and third butterfly valves. The residual mass at the bottom is equivalent to the little erosion occurring along the pipeline walls, which prevents the impacts of erosion from being visible all the way down the pipeline.

Simulation (a) exhibits more erosion than (b) for the 30-degree rotation, which contradicts the findings in Waja et al. (2022), where it stated that the erosion rate increases as the size of the particle increases, but for 45- and 60-degrees the erosion rate for simulation (b) is higher, which is in line with the findings that, potentially due to the ductile nature of the material, the drag forces acting on the particles or due to larger particles have more inertia inside the valve, resulting in reduced erosion.

In both simulations, the first butterfly valve encountered a lower erosion level than the others, the difference being in the 30-degree rotation in simulation (b). This further implies that there could be other factors that exert a substantial influence on erosion, such as drag force and disturbance in the flow, which can result in variable erosion outcomes.

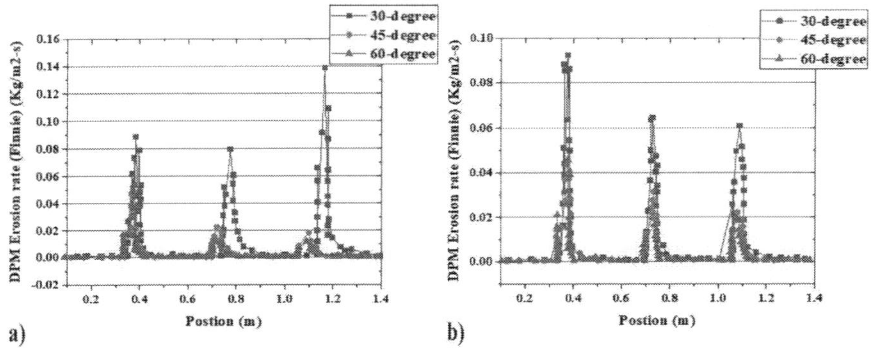

FIGURE 5.6 Erosion rate plot comparison.

Generally, as velocity increases, erosion increases, which is in line with the observations in Wei et al. (2020). The increase in fluid velocity magnitude on the valve areas, produced by the resistance resulting from a drop in the cross-sectional area when the valve shuts, leads to higher erosion rates at a valve opening angle of 30 degrees and lower rates at an angle of 60 degrees, which matches the following studies (Franco & Roberts, 1998; Fuyan & Hesheng, 1991), where it was shown that, for ductile materials, the rate of erosion peaks at around 20 to 30 degrees and then decreases at higher angles.

The static pressure increases when the opening angles are smaller; accordingly, the erosion is larger when the opening angle is smaller, since higher shear stresses at small opening angles might increase the erosion wear. As the opening angle decreases, the pressure drop increases, leading to energy dissipation that might result in decreased efficiency, heightened mechanical deterioration, and potentially increased erosion.

Parsi et al. (2014) find that the erosion of solid particulates is influenced by a variety of characteristics of the disintegrating substances and the materials being eroded, including size, shape, hardness, density, and material type, and Li et al. (2023) assert that the primary determinants of the profundity of erosion wear are the target material, usage, particle size, and velocity circumstances. However, they do not always guarantee higher erosion; other factors, such as impact angles, turbulence, drag forces, inertia, particle size, concentration, and even particle direction, can also play an essential role.

5.5 CONCLUSION AND FUTURE SCOPE

This research investigated the liquid flow in a pipeline with three butterfly valves, taking into account three distinct rotational angles and particle sizes.

The findings are as follows:

- A positive association between erosion rate and velocity magnitude indicates that an increase in erosion rate accompanies an increase in velocity magnitude on the valve areas.

- The chance of material fatigue and failure due to mechanical forces increases as the angle lowers since the pressure drop is greatest at that point.
- The rate of erosion rises with decreasing opening angle.
- Since the erosion pattern is less affected by particle size, flow dynamics seem to be the primary influencing factor.
- Because of the higher flow rate and potential turbulence, erosion occurs on the butterfly valve and its surroundings.
- When it comes to erosion, larger particles produce less than smaller ones.
- Due to its larger scale, simulation (b) produced a material erosion that was noticeably more dramatic, even if it had less erosion overall.
- Erosion may also be produced by several other factors, including impact angles, turbulence, drag forces, inertia, particle size, concentration, and even orientation.
- In general, a larger rotation angle indicates greater quality since it has less erosion and damage.

The research emphasizes the need to modify valve angles to lessen the effects of erosion, high pressure, and high velocity, extending the lifetime of valves and potentially resulting in cost savings and improved infrastructure. Furthermore, the study offers a strong framework for further investigation and advancement in fluid transport systems.

Future studies should focus on:

- Verifying this CFD experiment by executing a real-life experiment.
- Conducting an erosion wear and fluid flow analysis under a multitude of variables, including pressure, velocities, fluids, particle size, turbulence, and flow rates.
- The examination of multiphase flow, including gas–liquid and liquid–solid flows, which may offer a novel viewpoint on the investigation of erosion.
- The investigation of erosion and fluid flow through the use of multiple or distinct methodologies in CFD, such as DEM or erosion–corrosion.
- Future research, which may concentrate on pipeline coatings and high-temperature degradation materials.

List of Abbreviations

7D: seven times the diameter
ANN: artificial neural networks
CFD: computational fluid dynamics
DEM: discrete element method
DPM: discrete phase model
j/kg-k: joule per kilogram per kelvin
kg/m^2-s: kilogram per meter square second
kg/m^3: kilogram per cubic meter
kg/m-s: kilogram per meter second
LNG: liquified natural gas

m/s: meters/second
m/s^2: meters per second squared
m: meter
mm: millimeter
Pa: Pascal
RANS: Reynolds averaged Navier–Stokes
SST: shear stress transport
w/m-k: watt above meter kelvin

REFERENCES

Del Toro, A., Johnson, M. C., & Spall, R. E. (2015). Computational fluid dynamics investigation of butterfly valve performance factors. *Journal—American Water Works Association*, *107*(5), E243–E254. American Water Works Association. https://doi.org/10.5942/jawwa.2015.107.0052

Duan, Y., Eaton, M. D., Bluck, M. J., & Jackson, C. (2018). Assessments of different turbulence models in predicting the performance of a butterfly valve. *International Conference on Nuclear Engineering, Proceedings, ICONE*, *8*. https://doi.org/10.1115/ICONE26-82376

Elaswad, R., El-Sheimy, N., & Mohamad, A. (2024). Trajectory dynamics of particles in accelerated toroidal pipe: A computational study using CFD-DPM simulations. *Journal of Computational Science*, *78*. https://doi.org/10.1016/j.jocs.2024.102285

Fernandes, C., Semyonov, D., Ferrás, L. L., & Nóbrega, J. M. (2018). Validation of the CFD-DPM solver DPMFoam in OpenFOAM ® through analytical, numerical and experimental comparisons. *Granular Matter*, *20*(4), 1–18. https://doi.org/10.1007/S10035-018-0834-X/METRICS

Finnie, I. (1960). Erosion of surfaces by solid particles. *Wear*, *3*(2), 87–103. https://doi.org/10.1016/0043-1648(60)90055-7

Finnie, I., & Kabil, Y. H. (1965). On the formation of surface ripples during erosion. *Wear*, *8*(1), 60–69. https://doi.org/10.1016/0043-1648(65)90251-6

Franco, A., & Roberts, S. G. (1998). The effect of impact angle on the erosion rate of polycrystalline α-Al2O3. *Journal of the European Ceramic Society*, *18*(3), 269–274. https://doi.org/10.1016/S0955-2219(97)00121-0

Fuyan, L., & Hesheng, S. (1991). The effect of impingement angle on slurry erosion. *Wear*, *141*(2), 279–289. https://doi.org/10.1016/0043-1648(91)90274-X

Gella, D., Maza, D., & Zuriguel, I. (2017). Role of particle size in the kinematic properties of silo flow. *Physical Review E*, *95*(5), 052904. https://doi.org/10.1103/PHYSREVE.95.052904/FIGURES/11/MEDIUM

Guan Song, X., & Park, Y. C. (n.d.). *Numerical Analysis of Butterfly Valve-Prediction of Flow Coefficient and Hydrodynamic Torque Coefficient.*

Hankeln, F., & Riedelbauch, S. (2019). Comparison of different models for determination of erosion wear in centrifugal pumps. *www.Euroturbo.Eu//*. https://doi.org/10.29008/ETC2019-037

Hee Lee, G., & Ho Bae, J. (n.d.). *Transactions of the Korean Nuclear Society Virtual Spring Meeting CFD Simulation of the Hydraulic Characteristics Inside a Butterfly Valve Using ANSYS CFX.*

Hu, X., Barker, R., Neville, A., & Gnanavelu, A. (2011). Case study on erosion-corrosion degradation of pipework located on an offshore oil and gas facility. *Wear*, *271*(9–10), 1295–1301. https://doi.org/10.1016/J.WEAR.2011.01.036

Huang, C., & Kim, R. H. (1996). Three-dimensional analysis of partially open butterfly valve flows. *Journal of Fluids Engineering*, *118*(3), 562–568. https://doi.org/10.1115/1.2817795

Janoszek, T., & Rotkegel, M. (123 C.E.). Coupled CFD-FEM analysis of the damage causes of the retention bunker: A case study at hard coal mine. *Scientific Reports*, *14*, 14189. https://doi.org/10.1038/s41598-024-65034-z

Karimi, S., Shirazi, S. A., & McLaury, B. S. (2017). Predicting fine particle erosion utilizing computational fluid dynamics. *Wear*, *376–377*, 1130–1137. https://doi.org/10.1016/J.WEAR.2016.11.022

Kyriakopoulos, G. L., Aminpour, Y., Yamini, O. A. Movahedi, A., Mousavi, S. H., & Kavianpour, M. R. (2022). Hydraulic performance of howell-bunger and butterfly valves used for bottom outlet in large dams under flood hazards. *Applied Sciences 2022*, *12*(21), 10971. https://doi.org/10.3390/APP122110971

Leskovec, M., Lundell, F., & Innings, F. (2020). Pipe flow with large particles and their impact on the transition to turbulence. *Physical Review Fluids*, *5*(11). https://doi.org/10.1103/PHYSREVFLUIDS.5.112301

Li, S., Zhang, B., Yang, L., Zhang, J., Wang, Y. & Kang, W. (2023). Study on wear properties of the graphite-sealing surfaces in a triple eccentric butterfly valve based on EDEM-fluent coupling. *Machines*, *11*(4). https://doi.org/10.3390/machines11040463

Liu, B., Zhao, J., & Qian, J. (2017). Numerical study of solid particle erosion in butterfly valve. *IOP Conference Series: Materials Science and Engineering*, *220*(1). https://doi.org/10.1088/1757-899X/220/1/012018

Luaibi, M. S., & Abdulwahid, M. A. (2022). Numerical analysis by computational fluid dynamic simulation of fluid flow in a T-junction. https://doi.org/10.4108/eai.7-9-2021.2314888

Mohebi, A. R., Najafi, F., Mozaffarian, M., Dabir, B., & Esmaeilian Amrabadi, N. (2024). Fluid dynamics and erosion analysis in industrial naphtha reforming: A CFD-DPM simulation approach. *Powder Technology*, *436*, 119417. https://doi.org/10.1016/J.POWTEC.2024.119417

Nystad, B. H., Gola, G., Hulsund, J., & Roverso, D. (2010). *Technical Condition Assessment and Remaining Useful Life Estimation of Choke Valves subject to Erosion.*

ORNL/NRC/LTR-91/25, "Throttled Valve Cavitation and Erosion". (1991).

Pandya, D. A., Dennis, B. H., & Russell, R. D. (2017). A computational fluid dynamics based artificial neural network model to predict solid particle erosion. *Wear*, *378–379*, 198–210. https://doi.org/10.1016/J.WEAR.2017.02.028

Parsi, M., Najmi, K., Najafifard, F., Hassani, S., McLaury, B. S., & Shirazi, S. A. (2014). A comprehensive review of solid particle erosion modeling for oil and gas wells and pipelines applications. *Journal of Natural Gas Science and Engineering*, *21*, 850–873. https://doi.org/10.1016/J.JNGSE.2014.10.001

Perera, P., Hayward, K., Guzzomi, F., & Vafadar, A. (2024). Erosion wear characterisation of an open Ductile Iron butterfly valve subjected to Aluminium Oxide particle slurry flow. *Tribology International*, *191*, 109199. https://doi.org/10.1016/J.TRIBOINT.2023.109199

Song, X. (2007). *Numerical Analysis of Butterfly Valve-Prediction of Flow Coefficient and Hydrodynamic Torque Coefficient.*

Sotoodeh, K. (2023). The mathematical analysis and review of water hammering in check valves in offshore industry. *Journal of the Institution of Engineers (India): Series C*, *104*(4), 879–885. https://doi.org/10.1007/S40032-023-00965-6/FIGURES/4

Stack, M. M., & Abdulrahman, G. H. (2012). Mapping erosion-corrosion of carbon steel in oil-water solutions: Effects of velocity and applied potential. *Wear*, *274–275*, 401–413. https://doi.org/10.1016/J.WEAR.2011.10.008

Waja, E. A. S., Kholi, F. K., Esu-Ejemot Aquah, G., Tulashie, S. K., & Nyasapoh, M. A. (2022). Parametric investigation of sand erosion rates in gas pipeline bends under typical industry conditions using simplified method. *Energy Sources, Part A: Recovery, Utilization, and Environmental Effects*, *44*(4), 9325–9341. https://doi.org/10.1080/15567036.2022.2129873

Wang, C., Neville, A., Ramachandran, S., & Jovancicevic, V. (2005). Alleviation of erosion-corrosion damage by liquid-sand impact through use of chemicals. *Wear*, *258*(1–4), 649–658. https://doi.org/10.1016/J.WEAR.2004.09.038

Wei, L., Pei, L., & Chen-Xu, S. (2020). *Simulation Prediction of 60.3 mm Elbow Erosion Based CFD*. https://api.semanticscholar.org/CorpusID:222071820

Wojtkowiak, J., & Oleśkowicz-Popiel, C. (2006). *Investigations of Butterfly Control Valve Flow Characteristics*. https://www.researchgate.net/publication/242672823

Xu, B., Zhu, Z., Lin, Z., & Wang, D. (n.d.a). Solid-liquid two-phase flow and erosion calculation of butterfly valves at small opening based on DEM method. https://doi.org/10.1108/ILT-07-2020-0264/Keywords

Xu, B., Zhu, Z., Lin, Z., & Wang, D. (n.d.b). Solid-liquid two-phase flow and erosion calculation of butterfly valves at small opening based on DEM method. https://doi.org/10.1108/ILT-07-2020-0264/Keywords

Zhang, G., Wang, W. W., Zhang, H. T., Kim, H. D., & Lin, Z. (2023). Characteristics of cavitation evolution through a butterfly valve under transient regulation. *Physics of Fluids*, *35*(2). https://doi.org/10.1063/5.0137019

Zhang, M., Zhang, Q., Wang, X., Zhou, Z., & Liao, R. (2022). Study on the erosion wear in pipe fittings. *Journal of Engineering Research (Kuwait)*, *10*(1B), 138–152. https://doi.org/10.36909/jer.8372

6 Enhancing Concrete Performance with Sugarcane Bagasse Ash

Impact on Workability, Strength, and Durability

Mohammad Tabish, Mohd Moonis Zaheer, Abdul Baqi, and Salim Barbhuiya

6.1 INTRODUCTION

Concrete, a common construction material, is generally composed of cement, water, and aggregates. Nonetheless, the production of cement plays a crucial role in the global landscape of carbon dioxide emissions, accounting for approximately 7–8% of the total CO_2 emissions worldwide (Barbhuiya et al., 2020; Tabish et al., 2023). The construction industry is increasingly motivated to find sustainable alternatives to traditional Portland cement, largely by incorporating supplementary cementitious materials (SCMs). A notable sustainable construction material is sugarcane bagasse ash (SCBA), which is a by-product of the sugar industry. In agricultural nations such as India, significant amounts of agro-waste, including sugarcane bagasse, are produced. The fibrous residue remaining after juice extraction from sugarcane is known as sugarcane bagasse, and when this residue is burned, it produces ash that serves as a valuable resource for sustainable concrete production. Traditionally, this waste has been disposed of by burning or landfilling, leading to environmental pollution. The high amorphous silica content in SCBA makes it an ideal pozzolanic material, suitable for partially replacing cement in concrete production. Its pozzolanic properties allow SCBA to react with calcium hydroxide in concrete, enhancing strength and durability while contributing to more sustainable construction practices. (Minnu et al., 2021; Zaheer & Tabish, 2023).

The utilization of SCBA not only addresses waste management issues but also contributes to the reduction of cement usage, thereby lowering the carbon footprint of concrete. Incorporating SCBA into concrete can improve its compressive strength, tensile strength, durability, and permeability resistance. The improvements arise from pozzolanic reactions involving the amorphous silica present in SCBA and calcium hydroxide found in the cement paste, resulting in additional calcium silicate

DOI: 10.1201/9781003502647-7

hydrate (C–S–H) gel formation, thereby reinforcing and compacting the concrete matrix. India, being a leading producer of sugarcane, produces significant quantities of bagasse, which serves as a plentiful source of SCBA for producing sustainable concrete. Annually, the country produces about 300 million tonnes of sugarcane, leading to approximately 10 million tonnes of bagasse ash. This abundant availability of SCBA presents a tremendous opportunity to utilize this agro-waste effectively in concrete production (Murugesan et al., 2022). Integrating SCBA into concrete offers India an opportunity to decrease reliance on traditional cement, advance sustainable construction, and lessen the environmental impact of cement production and agricultural waste disposal. A key factor in optimizing SCBA's effectiveness as a pozzolanic material is particle size. Reducing SCBA particle size through fine grinding significantly enhances its pozzolanic reactivity by increasing surface area, which promotes more intensive reactions with calcium hydroxide in the cement paste, resulting in stronger, more durable concrete. Ultrafine SCBA particles provide a better filler effect, filling the voids in the concrete matrix and leading to a denser, more homogeneous material. This enhances the mechanical properties of concrete while also improving its durability and resistance to environmental degradation. Studies have shown that the particle size reduction of SCBA can result in superior pozzolanic performance. For instance, grinding SCBA to a mean particle size of around 17 microns has been found to significantly increase its reactivity. The increased surface area enables more effective interaction with the cement matrix, encouraging greater formation of C–S–H gel and boosting the concrete's overall strength and durability. This results in improved performance across key structural and durability parameters. This makes finely ground SCBA a highly effective SCM for improving the properties of concrete while also addressing environmental concerns associated with cement production and agricultural waste disposal (Athira & Bahurudeen, 2022; Bahurudeen et al., 2016).

This research aims to utilize SCBA as a partial cement replacement in concrete, addressing environmental impacts and waste management. Concrete specimens with 10%, 15%, 20%, and 25% SCBA were produced. In agricultural countries like India, using SCBA promotes sustainable construction, enhances concrete performance, and reduces the construction industry's carbon footprint. This research seeks to examine how SCBA particle size influences its pozzolanic activity, aiming to maximize its effectiveness as a cement substitute in concrete.

6.2 MATERIALS AND METHODOLOGY

6.2.1 Materials

Grade 43 Ordinary Portland Cement (OPC 43), conforming to BIS 8112: 1989 standards, was utilized as the binder in this study. Natural river sand served as the fine aggregate, while coarse aggregate was also employed. The SCBA was processed in a ball mill with a 5 kg capacity, using steel balls. This milling process involved 9000 revolutions, which effectively reduced the average particle size to 17.24 microns. Table 6.1 outlines the physical properties of the cement and SCBA, while Table 6.2 provides a comprehensive overview of the physical properties of the river sand and

TABLE 6.1
Physical Properties of OPC and SCBA

Properties	Material	
	OPC	**SCBA**
Colour	Grey	Black
Specific gravity	3.01	1.87
Standard consistency (%)	33	49
Initial setting time (minutes)	40	60
Final setting time (minutes)	220	345

TABLE 6.2
Physical Properties of River Sand and Coarse Aggregate

Properties	Material	
	Sand	**Coarse Aggregate**
Specific gravity	2.66	2.61
Water absorption (%)	1.08	1.95
Fineness modulus	2.97	6.73

coarse aggregates. The particle size distribution of the grounded SCBA is illustrated in Figure 6.1, while the scanning electron microscope (SEM) image is presented in Figure 6.2.

6.2.2 Concrete Mix Design

The study involved the partial replacement of cement with ground SCBA at varying levels of 10%, 15%, 20%, and 25% by weight. Maintaining a constant water-to-cement ratio of 0.478, the incorporation of SCBA resulted in an increased water demand attributed to its extensive specific surface area and higher carbon content. To achieve sufficient workability, a polycarboxylic ether-based superplasticizer, meeting ASTM C494 Type-F standards, was added to the mix. The mix proportions utilized in this study are detailed in Table 6.3.

6.2.3 Methods

The workability of the fresh concrete mix was assessed according to IS 1199 standards. For hardened properties, compressive strength tests were carried out on 150 mm cubes at 7 and 28 days, following IS 516:1959 guidelines. Ultrasonic pulse velocity (UPV) measurements were performed on these cubes using a 54 Hz transducer, in accordance with ASTM C597. The evaluation of split tensile strength involved cylindrical specimens measuring 150 mm in diameter and 300 mm in length. These

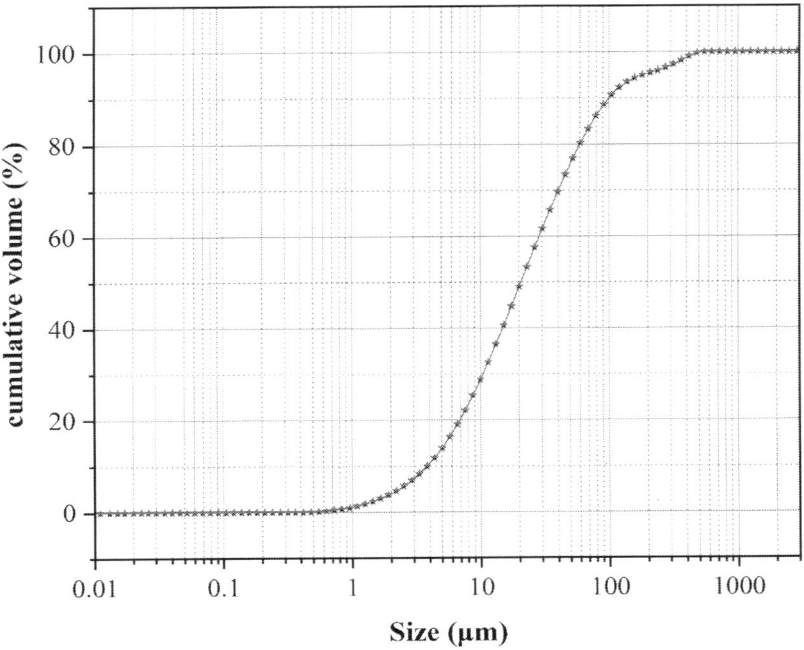

FIGURE 6.1 Particle size distribution of grounded SCBA.

FIGURE 6.2 SEM of grounded SCBA.

TABLE 6.3
Mix Proportions

Mix	Cement (kg/m³)	SCBA (kg/m³)	Fine Aggregate (kg/m³)	Coarse Aggregate (kg/m³)		Water (kg/m³)	Superplasticizer (kg/m³)
				10 mm	20 mm		
CM	349	0	711	488.4	732.6	167	3.49
S10	314.1	34.9	711	488.4	732.6	167	4.36
S15	296.65	52.35	711	488.4	732.6	167	5.23
S20	279.2	69.8	711	488.4	732.6	167	6.10
S25	261.75	87.25	711	488.4	732.6	167	6.98

specimens were also utilized to assess surface resistivity using a PROCEQ Resipod device. This device, equipped with four 50 mm-spaced probes, measured surface resistivity on specimens cured for 7 and 28 days. Water absorption testing was conducted on a cube with a dimension of 150 mm, cured in water for 7 and 28 days, following ASTM C642 standards. Specimens were demoulded 24 ± 1 hours after casting and subsequently cured in a water tank maintained at 20°C ± 1°C until the specified testing ages. Several concrete mixes were prepared for this experiment. The control mix (CM), without SCBA, was designated as CM. The other mixes incorporated SCBA as a cement replacement by weight at levels of 10%, 15%, 20%, and 25% and were designated as S10, S15, S20, and S25, respectively.

6.3 RESULTS AND DISCUSSIONS

6.3.1 Workability

The workability of concrete mixes was evaluated by the slump test. Figure 6.3 illustrates the effect of SCBA on slump values, demonstrating a distinct reduction in slump as SCBA content increases. The decrease can be linked to the elevated carbon content and particular surface area (SSA) of SCBA particles. The unburnt, porous, and fibrous carbon particles in SCBA exhibit a higher water absorption rate compared to cement particles during mixing. This leads to a decrease in the amount of free water available for particle lubrication, which in turn results in lower slump values as the replacement of SCBA increases. Additionally, the higher SSA of ground SCBA increases the water demand. These factors collectively contribute to the observed decrease in slump values. These findings are consistent with previous studies (Bahurudeen et al., 2014; Chagas et al., 2009; Chusilp et al., 2009), corroborating the effect of SCBA on concrete workability.

6.3.2 Compressive Strength

The compressive strength of concrete mixes was evaluated after 7 and 28 days of curing, with the findings presented in Figure 6.4. The findings indicate that the

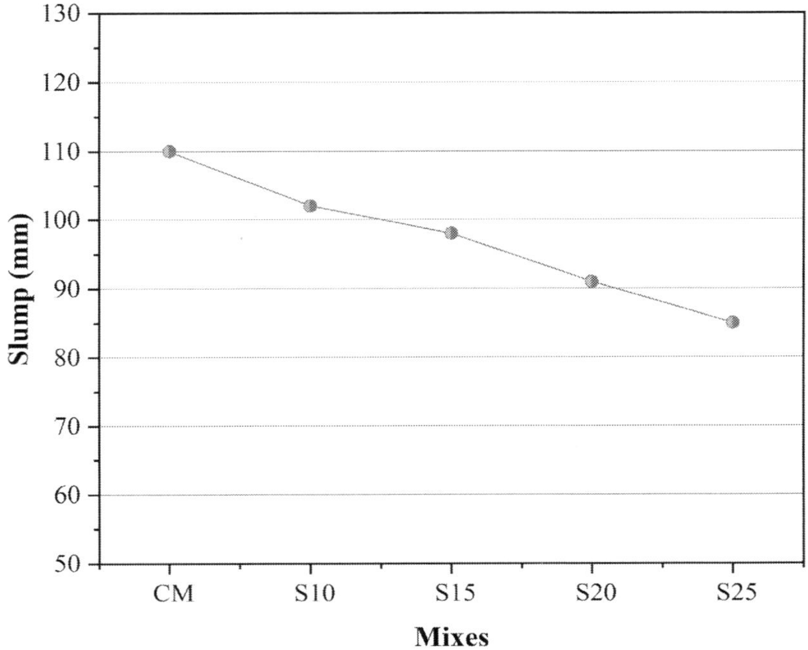

FIGURE 6.3 Slump values of different mixes.

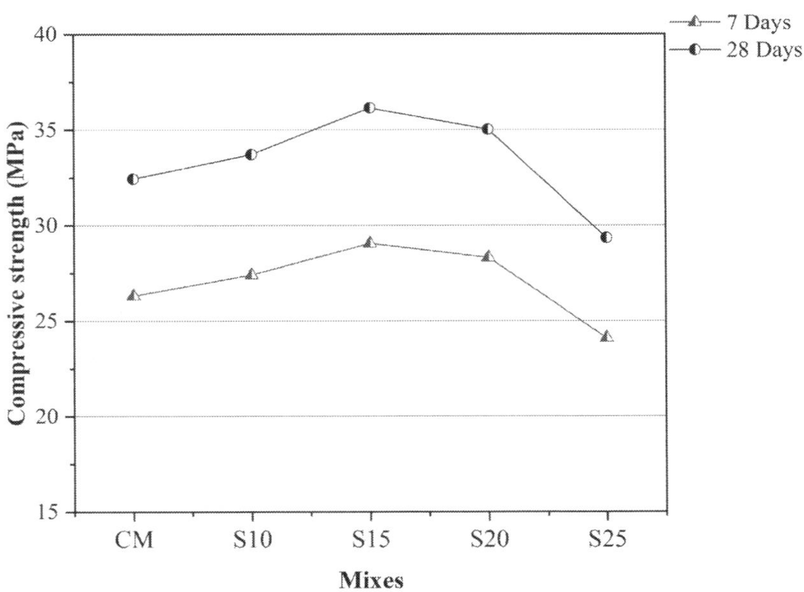

FIGURE 6.4 Compressive strength of all mixes at 7 and 28 days.

compressive strength improves with SCBA replacement, reaching a maximum at 20%, exceeding the performance of the CM. Nonetheless, with a 25% replacement, there is a reduction in compressive strength. The combination containing 15% SCBA (S15) demonstrated the greatest compressive strength at both the 7- and 28-day marks. At seven days, S10, S15, and S20 showed increases of 4.18%, 10.37%, and 7.56%, respectively, over the CM's 26.32 MPa strength, while S25 experienced an 8.42% decrease.

After 28 days, S10, S15, and S20 showed compressive strength increases of 3.91%, 11.43%, and 7.95%, respectively, over the CM at 32.44 MPa, whereas S25 showed a 9.56% decrease. The observed increase in strength can be linked to the filler effect and pozzolanic reactivity of SCBA. In this process, the reactive silica present in SCBA interacts with calcium hydroxide, resulting in the formation of additional C–S–H, which enhances the matrix's strength. These results align with previous studies (Bahurudeen et al., 2015; Rajasekar et al., 2018).

6.3.3 Split Tensile Strength

The results for the splitting tensile strength of the concrete mixes are presented in Figure 6.5. The findings show that the splitting tensile strength first rises with the addition of SCBA but then decreases at elevated replacement levels. The S15 mix demonstrated the greatest splitting tensile strength, whereas the S20 mix also

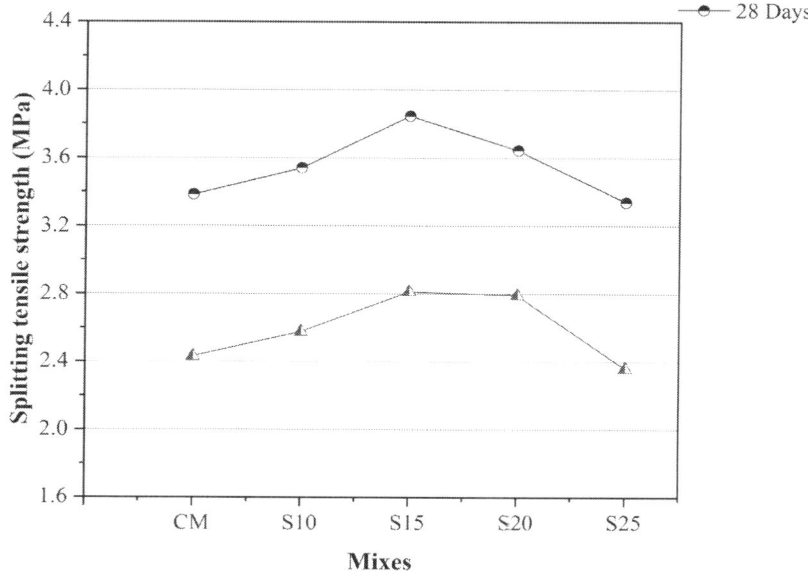

FIGURE 6.5 Splitting tensile strength of all mixes at 7 and 28 days.

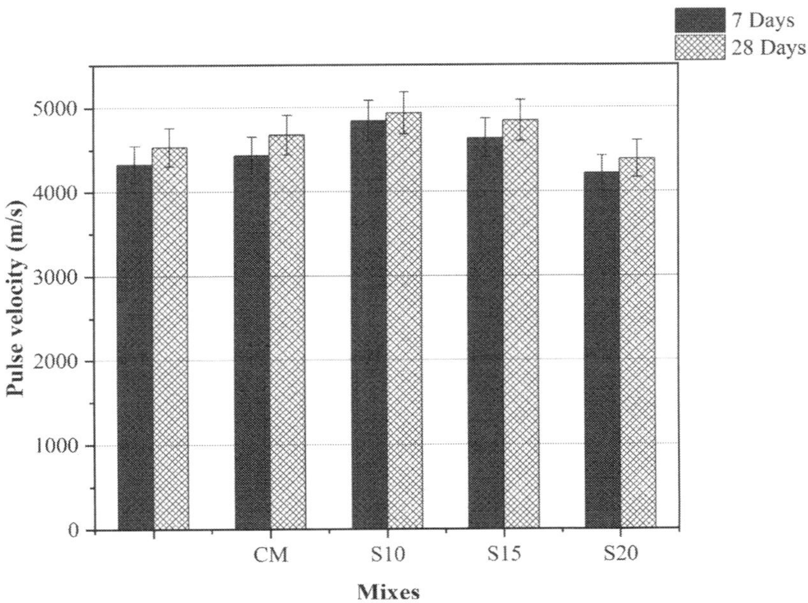

FIGURE 6.6 Effect of SCBA on UPV at 7 and 28 days.

displayed enhanced strength in comparison to the CM. The tensile strength of concrete mixes incorporating SCBA replacements of 10%, 15%, and 20% (S10, S15, and S20) exhibited increases of approximately 5.96%, 15.70%, and 14.84%, respectively, when compared to the CM. In contrast, the S25 mix demonstrated a minor reduction in strength of 2.96%.

After 28 days, the splitting tensile strength of the S10, S15, and S20 mixes was 4.64%, 13.65%, and 7.68% higher, respectively, than that of the CM. However, the S25 mix demonstrated a marginal strength decrease of 1.27%. These findings indicate that moderate SCBA replacement enhances splitting tensile strength, but excessive replacement can reduce performance. The enhancement in strength results from the pozzolanic reaction occurring between SCBA and calcium hydroxide during the hydration of cement, leading to the formation of additional C–S–H gel. This process enhances binding and matrix density, as corroborated by earlier research (Bahurudeen et al., 2015).

6.3.4 UPV Test

The UPV results are instrumental in determining the homogeneity of concrete. As shown in Figure 6.6, the UPV values initially increase with the addition of SCBA, reaching a peak before starting to decline. This increase in UPV can be attributed to the ultrafine particles of ground SCBA, which enhance the concrete's filler effect and promote additional C–S–H gel formation through pozzolanic reactions

(Hajimohammadi et al., 2018; Liu et al., 2014). This process results in a denser cement matrix and an improved microstructure, allowing ultrasonic waves to travel more quickly to the receiver transducer, thereby reducing transit time and increasing UPV values. After seven days of curing, the UPV values of the S10, S15, and S20 mixes were 2.37%, 11.77%, and 7.04% higher, respectively, compared to CM. After 28 days of curing, these increases were 3.13%, 8.80%, and 6.83%, respectively, relative to the CM. The UPV values of S25 mixes at 7 and 28 days were reduced by 2.61% and 3.22% compared to CM. The results indicate that a moderate replacement of SCBA improves the homogeneity and microstructural integrity of concrete, whereas an excessive amount of SCBA could result in a deterioration of these properties.

6.3.5 Electrical Resistivity

Figure 6.7 illustrates the surface resistivity of concrete mixes at 7 and 28 days. The resistivity of SCBA-incorporated mixes showed a significant increase compared to the CM, falling within the range of moderate to low chloride permeability. At seven days, the S15 mix exhibited the highest resistivity value of 21 kΩ-cm, which was 105.8% higher than the CM. By 28 days, the S20 mix displayed the highest resistivity value. The resistivity values for the S10, S15, S20, and S25 mixes were 18.34%, 41.72%, 63.29%, and 8.23% higher, respectively, than the CM. All samples

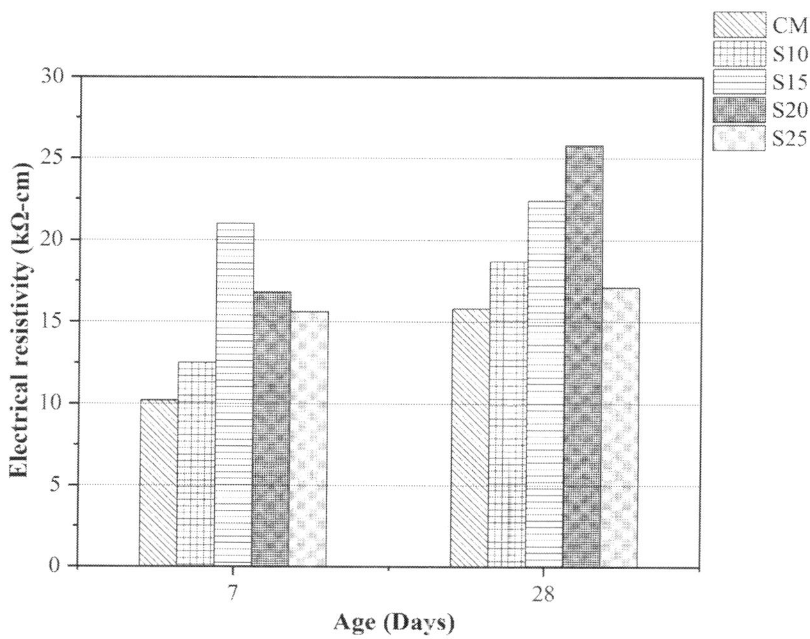

FIGURE 6.7 Surface resistivity of different mixes at 7 and 28 days.

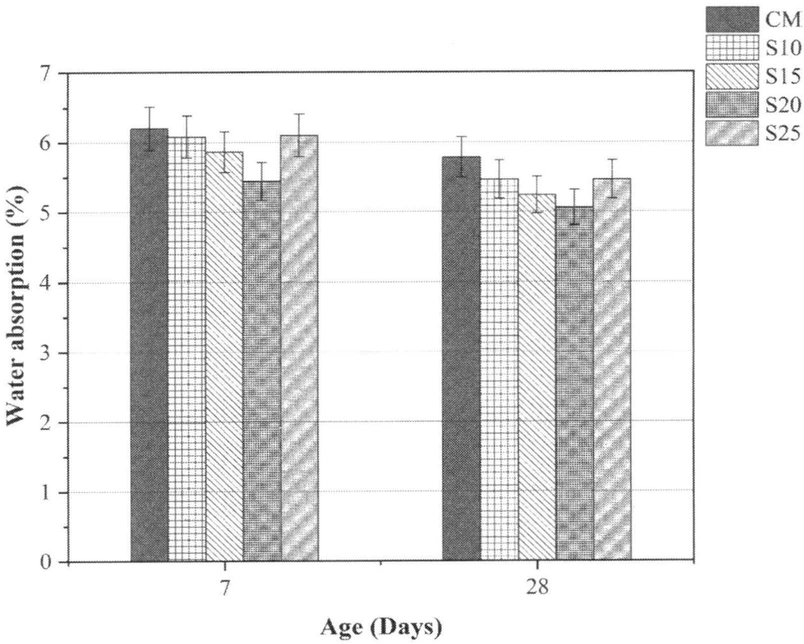

FIGURE 6.8 Effect of SCBA on water absorption at 7 and 28 days.

demonstrated an increase in resistivity over time, particularly during the initial phase. This significant increase is attributed to the ongoing hydration and hardening processes, which reduce the porosity of the mixes and result in higher resistivity values. The ultrafine particles of SCBA contribute to a denser cement matrix, further enhancing the resistivity (Joshaghani & Moeini, 2018). The results indicate that the inclusion of SCBA improves not only resistivity but also durability and resistance to chloride permeability in the concrete mixtures.

6.3.6 Water Absorption

Figure 6.8 shows the results of water absorption for concrete mixes. The data show a clear decrease in water absorption with the addition of SCBA, along with a reduction over time. At seven days, water absorption rates for S10, S15, S20, and S25 mixes were 6.08%, 5.86%, 5.44%, and 6.10%, respectively, compared to 6.20% for the CM. By 28 days, water absorption further declined for all SCBA mixes: S10, S15, S20, and S25 recorded rates of 5.46%, 5.24%, 5.06%, and 5.46%, respectively, while the CM was 5.78%. This reduction is attributed to matrix densification from the filler effect and pozzolanic reactivity of SCBA, which refines the concrete microstructure, lowers porosity, and decreases water absorption (Gupta et al., 2021; Mulay et al., 2016).

6.4 CONCLUSIONS

The findings of this study comparing CM concrete with SCBA-incorporated concrete can be concluded as follows:

- Incorporating SCBA in concrete reduces slump due to its high carbon content, SSA, and water absorption, negatively affecting workability.
- The compressive and split tensile strength of concrete mixes improved with the replacement of cement by ultrafine SCBA particles, with the optimal replacement level identified as 20%.
- Moderate SCBA replacement enhances concrete homogeneity and microstructure integrity, improving UPV values and indicating potential durability benefits.
- The incorporation of SCBA significantly improves the resistivity of concrete across all levels of replacement.
- The use of SCBA leads to a decrease in water absorption in concrete, which can be linked to the densification of the matrix resulting from the filler effect and pozzolanic activity, thereby improving its microstructure and durability.

REFERENCES

Athira, G., Bahurudeen, A., 2022. Rheological properties of cement paste blended with sugarcane bagasse ash and rice straw ash. Constr. Build. Mater. 332, 127377. https://doi.org/10.1016/j.conbuildmat.2022.127377

Bahurudeen, A., Kanraj, D., Gokul Dev, V., Santhanam, M., 2015. Performance evaluation of sugarcane bagasse ash blended cement in concrete. Cem. Concr. Compos. 59, 77–88. https://doi.org/10.1016/j.cemconcomp.2015.03.004

Bahurudeen, A., Marckson, A. V., Kishore, A., Santhanam, M., 2014. Development of sugarcane bagasse ash based Portland pozzolana cement and evaluation of compatibility with superplasticizers. Constr. Build. Mater. 68, 465–475. https://doi.org/10.1016/j.conbuildmat.2014.07.013

Bahurudeen, A., Wani, K., Basit, M. A., Santhanam, M., 2016. Assesment of pozzolanic performance of sugarcane bagasse ash. J. Mater. Civ. Eng. 28, 04015095. https://doi.org/10.1061/(asce)mt.1943-5533.0001361

Barbhuiya, G. H., Moiz, M. A., Hasan, S. D., Zaheer, M. M., 2020. Effects of the nanosilica addition on cement concrete: A review. Mater. Today Proc. 32, 560–566. https://doi.org/10.1016/j.matpr.2020.02.143

Chagas, G., Dias, R., Filho, T., Marcelo, L., Moraes, E. De, Fairbairn, R., 2009. Cement and concrete research ultra fine grinding of sugar cane bagasse ash for application as pozzolanic admixture in concrete. Cem. Concr. Res. 39, 110–115. https://doi.org/10.1016/j.cemconres.2008.11.005

Chusilp, N., Jaturapitakkul, C., Kiattikomol, K., 2009. Utilization of bagasse ash as a pozzolanic material in concrete. Constr. Build. Mater. 23, 3352–3358. https://doi.org/10.1016/j.conbuildmat.2009.06.030

Gupta, P., Wirquin, E., Bokhoree, C., 2021. Case studies in construction materials sustainable concrete: Potency of sugarcane bagasse ash as a cementitious material in the construction industry. Case Stud. Constr. Mater. 14, e00545. https://doi.org/10.1016/j.cscm.2021.e00545

Hajimohammadi, A., Ngo, T., Mendis, P., 2018. Enhancing the strength of pre-made foams for foam concrete applications. Cem. Concr. Compos. 87, 164–171. https://doi.org/10.1016/j.cemconcomp.2017.12.014

Joshaghani, A., Moeini, M. A., 2018. Evaluating the effects of sugarcane-bagasse ash and rice-husk ash on the mechanical and durability properties of mortar. J. Mater. Civ. Eng. 30, 04018144. https://doi.org/10.1061/(asce)mt.1943-5533.0002317

Liu, M. Y. J., Alengaram, U. J., Jumaat, M. Z., Mo, K. H., 2014. Evaluation of thermal conductivity, mechanical and transport properties of lightweight aggregate foamed geopolymer concrete. Energy Build. 72, 238–245. https://doi.org/10.1016/j.enbuild.2013.12.029

Minnu, S. N., Bahurudeen, A., Athira, G., 2021. Comparison of sugarcane bagasse ash with fly ash and slag: An approach towards industrial acceptance of sugar industry waste in cleaner production of cement. J. Clean. Prod. 285, 124836. https://doi.org/10.1016/j.jclepro.2020.124836

Mulay, S., Vesmawala, G., Patil, Y., Gholap, V., 2016. Experimental investigation of sugarcane bagasse ash concrete under sodium hydroxide solution. Am. J. Civ. Eng. 5, 1–8. https://doi.org/10.11648/j.ajce.20170501.11

Murugesan, T., Vidjeapriya, R., Bahurudeen, A., 2022. Reuse of silica rich sugarcane bagasse ash in concrete and influence of different curing on the performance of concrete. Silicon 14, 3069–3080. https://doi.org/10.1007/s12633-021-01089-1

Rajasekar, A., Arunachalam, K., Kottaisamy, M., Saraswathy, V., 2018. Durability characteristics of ultra high strength concrete with treated sugarcane bagasse ash. Constr. Build. Mater. 171, 350–356. https://doi.org/10.1016/j.conbuildmat.2018.03.140

Tabish, M., Zaheer, M. M., Baqi, A., 2023. Effect of nano-silica on mechanical, microstructural and durability properties of cement-based materials: A review. J. Build. Eng. 65, 105676. https://doi.org/10.1016/j.jobe.2022.105676

Zaheer, M. M., Tabish, M., 2023. The durability of concrete made up of Sugar Cane Bagasse Ash (SCBA) as a partial replacement of cement: A review. Arab. J. Sci. Eng. 48, 4195–4225. https://doi.org/10.1007/s13369-023-07698-9

Section II

Transformation of Advanced Materials Processing Toward Industrial Engineering

7 Technological Prospecting for the Sugarcane Juice Extraction Stage

A Patentometric Study on the Diffuser and Crushing Mill Process

Fabrício José Piacente, Silva Vanessa de Cillos, Biaggi Denys Eduardo, Alexandre Formigoni, Rosinei Batista Ribeiro, Eliacy Cavalcanti Lélis, and Marco Aurélio Feriotti

7.1 INTRODUCTION

According to data from the USDA (2022), world sugar production in the 2021/22 harvest was 180.3 million tons; for the 2021/22 harvest, small growth is expected (1.5%) and should reach 183.1 million tons. Increased production in Thailand, China and Russia is expected to be offset by reductions in India and the European Union. Brazil, the largest producer in the world, yielded 41.8 million tons, followed by India with 32.3 million tons and then Thailand with 10.2 million tons. In the 2021–2022 harvest, sugarcane occupied an approximate area of 28.5 million hectares; in Brazil 9.1 million hectares were planted, 5.1 million hectares in India, 1.9 million hectares in China and 1.3 million hectares in Thailand (FAOSTAT, 2022).

The sugar prices have recovered since March 2022 and have remained much more attractive than ethanol prices. Therefore, sugar–ethanol plants should remain focusing on sugar production, and the sugar–ethanol production mix will likely be similar to the previous season. The Brazilian exports in 2021/22 are 32.15 mmt, raw value, the largest volume exported by Brazil, representing roughly 50% of the world exports (USDA, 2022).

Historically, sugarcane has been planted in tropical regions for the purpose of producing sugar. In recent decades, sugarcane has also been used to produce ethanol

DOI: 10.1201/9781003502647-9

on a commercial scale, a biofuel that partly replaces hydrocarbon derivatives such as gasoline. Once harvested in the field, sugarcane is transported to an agro-industrial processing unit called a sugarcane factory. In this processing unit, the sugarcane is lavaged, chopped and crushed to extract the juice, basically composed of sucrose and water.

The sugarcane juice extraction process can occur in two ways: the traditional and most used method in the world is grinding in sets of cylinders. The second process, recently more used, is diffusion. It consists of removing sucrose from sugarcane through a chemical process called leaching. In this process, the sugarcane is chopped and washed with water at a controlled temperature until all the sucrose is removed from the plant. This is a technology based on chemical principles.

Technological prospecting aims to map the evolution of the scientific development of technologies that are being developed and that can provide significant improvements in different sectors of the economy. Thus, technological prospecting studies constitute the basic tool to guide efforts undertaken to develop technologies. They can be treated as subsidies to expand the capacity for anticipation and to stimulate innovation systems in different companies (Piacente et al., 2015).

Petentometry is one of the technological prospecting tools used for mapping, big data analysis and strategic use of technological information contained in patent documents deposited in industrial protection offices.

This chapter aimed to map, through a technological prospecting study that uses the patentometry tool, the recent technological development of the two main technologies used in the extraction of sucrose from sugarcane: milling and diffusion. The research was based on patent documents deposited between the years 1975 and 2021, in the databases of the main industrial protection offices in the world. The main purpose was to characterize the development of these two technologies, highlighting the number of documents deposited, the country of origin of the deposit, the main depositing institutions and the nature of these depositors.

7.2 THEORETICAL FOUNDATION

In Brazil and other countries, the production of sugarcane and its processing can be considered one of the most important agro-industrial activities. The Brazilian sugar–alcohol sector, currently called sugar-energy, is made up of companies that only produce ethanol, called distilleries; the factories that produce sugar and ethanol are called sugarcane mills. Both can produce energy through cogeneration. The name "industrial complex" covers other manufacturing activities and various services and others that operate in some link in the production chain of these elements (Piacente et al., 2021).

The sugarcane mill process is a technological paradigm based on the mechanical principle of separation by crushing. For this separation of the bagasse (solid part) and sucrose (liquid part) to occur, the chopped sugarcane is introduced into sets of mills made up of cylindrical metal rollers. According to Rein (2013), a classic mill assembly works with three grooved rollers, two installed in the same lower plane as the third, as shown in Figure 7.1. This triangular arrangement allows the raw material to be squeezed twice. Mills may have a fourth roller, called a pressure roller, which improves feeding efficiency, maintaining a constant flow of sugarcane into the system.

A medium or large sugarcane mill in Brazil works with a set of five to six mills interconnected in a line. During the passage of the bagasse from one mill to another, imbibition takes place, that is, the addition of water for the purpose of increasing sucrose extraction (Piacente et al., 2016).

According to Stupiello (1987), the efficiency of a set of mills can be measured by two parameters: capacity and extraction efficiency. The capacity of a grinding suit is understood to be the quantity of sugarcane ground per unit of time. The term "extraction efficiency" refers to the amount of sucrose extracted from sugarcane by mills. Studies carried out by Castro et al. (2018) in a set of six in-line mills showed average extraction rates of 96.5%, and the final moisture of the bagasse was around 50%. According to these authors, some factors affected the crushing capacity, including sugarcane preparation, mill feed efficiency and the size and type of mill cylinders.

Diffusion is a chemical process that consists of removing sucrose by washing (leaching) the sugarcane with hot water, as shown in Figure 7.2. This is the main technique originally used to extract sucrose in the beet sugar industry. In this process, the beet is initially chopped and placed on a perforated mat and is subsequently washed in several cycles with jets of hot water, within a closed hydraulic system. As the beet is washed, the sucrose concentration of the water jets increases until all the sucrose is removed from the raw material. This technology was improved and adapted to the physiological conditions of sugarcane (Breward et al., 2012).

According to Piacente et al. (2016), experiments carried out in Germany by beet diffuser manufacturers resulted in a new design of continuous diffusion, more economical, efficient and easy to operate and which was adapted to the needs of the sugarcane industry. Currently, the diffusers that process sugarcane are horizontal, closed devices, made up of a perforated conveyor belt where the chopped cane is deposited. As the sugarcane moves inside the diffuser, jets of hot water are released, washing the cane and removing the sucrose. The extraction of sucrose occurs by mass transfer between two liquid phases in different concentrations under the action of the absorption fluid. This process guarantees high yield in sucrose extraction.

For Cavalcanti (2005), the diffuser presents important comparative advantages in relation to the grinding system. Among them, this author highlights the following: (i) greater extraction of sucrose, with numbers that can reach more than 98%; (ii) lower investments in equipment and installations, an average saving of 10% to 15% in the initial investment in equipment is expected with the complete installation of the diffuser; (ii) reduced annual maintenance costs, around 30% of a traditional system with mills. And the main disadvantage of using diffusers is that they carry more impurities with the bagasse, requiring attention with the boilers.

According to Piacente et al. (2016), regardless of the sucrose extraction system adopted, it's a fact that different companies that make up the industry of machinery and equipment have been concerned over the years with developing or adapting their technology to the objectives of the sugarcane processing sector. That's why they develop technical solutions within their Research and Development (R&D) department, which are tested and approved by the market with the potential to become innovations. From this point on, these innovations are protected by a patent document.

FIGURE 7.1 Milling set for sugarcane processing.

(Piacente, 2010).

FIGURE 7.2 Diffuser for sugarcane processing.

(Piacente, 2010).

For Silva et al. (2021), a patent is a title granted by the state that grants the exclusive right to exploit the invention or technological improvement to right holders for a certain period. This temporary monopoly gives the patent holder the right to prevent third parties, without their consent, from producing, using, offering for sale, selling or importing their technological invention. In Brazil, there are two types of patent protection: an invention patent and a model patent of utility.

The temporary monopoly that an intellectual property system provides has the purpose of contributing, over time, so that the investment made in the new product can be recovered and reinvested in new technological efforts, which stimulates the innovative process. This is an important device to protect the inventive capacity of economic agents that differentiate themselves in the market from other agents through new products or new processes or through technological improvements to existing products (Macedo, 2000).

The purpose of a legal patent system is to encourage technological progress and invention by granting the inventor a concession for a temporary monopoly on his creation. This system, as we know it, was created based on international agreements and treaties that shaped its general operating lines. The base document for this was those established in the Paris Union Convention (CUP) in 1883, to which member countries are signatories.

Currently, patent databases contain thousands of documents deposited annually around the world and constitute an important collection for studying the evolution of technological advancement in certain areas of knowledge. These documents are organized based on a rule established in the International Patent Classification (IPC), which serves to index documents and thus assists in searching for patent documents. Patentometry is the careful analysis of patent documents. It is one of the most used technological prospecting tools when seeking a specific technological solution to a problem; diagnosing the state of the art of a given technology; studying technological trajectories; and raising indicators to support science and technology policies and others (Piacente et al., 2016).

Technological prospecting methods can be classified into three groups: (i) evaluation, which is the systematic and continuous monitoring of the evolution of facts and identification of change carriers; (ii) forecasting, which consists of making projections based on historical information and modeling trends; and (iii) prospective, which consists of prediction possibilities, based on unstructured interaction between experts (Piacente & Silva, 2015).

7.3 METHODOLOGY

To construct this chapter, the methodology used was structured in the mapping of information contained in patent documents related to the grinding and diffusion processes for the extraction of sucrose from sugarcane. Priority was given to documents deposited between 1975 and 2021, which were deposited in the main industrial protection offices in the world.

The prospecting tool that supported this patentometry study was the Questel Orbit software, which is a search and analysis system for information contained in patents and industrial designs. This software allows searches to be carried out using keywords, bibliographic fields, patent classification and previous citations, among others. Questel Orbit has a comprehensive search of patent offices in approximately 96 countries, including complete information and files available for download.

The search syntax carried out in this research through Questel Orbit used the following terms as key words: (i) diffusion process: "cane* or cane* or bagasse*" and "diffuser"; (ii) grinding process: "sugar cane or sugarcane" and "crushing* or milling*" and IPC Code=(C13* OR B30B* OR B02C*).

The IPC has as its initial objective the establishment of an effective search tool for the retrieval of patent documents by intellectual property offices and other users, to establish novelty and evaluate inventive activity of technical disclosures in patent applications. It is a system legally adopted in more than 100 countries and has the administrative prerogative of the World Intellectual Property Organization.

Specifically for the case in question, the IPC indicates that C13 corresponds to section C, referring to "Chemistry and Metallurgy" and Class C13 on the "Sugar Industry". Section B corresponds to technologies linked to "Processing and Transport Operations"; Class B02C refers to "Crushing, Pulverizing or Disintegrating; Preliminary Processing of the Grain before Grinding". The B30B classification refers to patent documents linked to "Magnetic or Electrostatic Separation of Solid Materials from Solid or Fluid Materials; Separation by Means of High-Voltage Electric Fields".

The objective was to search all patent documents found for these two technologies and analyze the evolution that occurred during the analysis period. In this way, together with other information, we were able to find (i) the number of patents per technological subdomain; (ii) the number of patents by country of origin of the holder; (iii) the number of patents per holder; and (iv) the current status of the researched patent documents.

7.4 RESULTS AND DISCUSSIONS

The search for terms described in the methodology was carried out using the Questel Orbit software. As a result, 287 patent documents related to crushing mills and 35 related to diffusers were observed, which illustrates greater interest and research in the crushing process than in the diffuser.

Figure 7.3 shows the results of the search terms in relation to the year of patent filing in the period from 1975 to 2021. It is observed, throughout the analyzed period, that patent deposits were found for the term crushing. In the years 1996, 2009, and 2019, there were ten documents deposited for this search term. The years 2010, 2013, 2014, 2015, 2016, 2020, and 2021 stood out for presenting more than ten documents deposited with this search term. In the period from 2006 to 2021, 163 patent documents were filed with the search term "crushing mill", which suggests the importance of the topic and relevance of research in the area.

This period of emphasis in terms of the number of patent filings with the search term "crushing mill" coincides with the period of expansion of sugarcane cultivation in Brazil, the world's largest producer. The technology of flexfuel cars, which run on both gasoline and ethanol, stands out for the period. It generated a stimulus for the sugarcane industry in Brazil, increasing demand for ethanol. This increase in ethanol consumption also caused an increase in production in other sugarcane-producing countries, such as India and Thailand. The growth in sugarcane and ethanol production generates investment in the sector and consequently an increase in research and technological innovations (Piacente et al., 2016).

As for the search term diffuser, 35 patent documents were deposited for the period analyzed (1975–2021). These documents are distributed over 20 years, focusing on 1978, 1986, 1994 and 2019, which presented at least three patent documents filed.

It can be seen in Figure 7.3 that, historically, science and technology efforts aimed at the stage of extracting sucrose from sugarcane are concentrated on the technological solution based on extraction by plants. Therefore, it is noted that the number of patent documents filed in the main offices in the world relating to the grinding extraction process is higher than those registered for diffusers.

The main justifications for the technological option of extraction by grinding instead of a diffuser are the cost of purchasing the equipment in its flexibility and adaptability to increases in production, the accumulated learning to use technology and the existence of a larger network of manufacturers and service providers that serve the plants than the diffusers.

During the period analyzed, more efficient alternatives in the grinding and diffusion processes were developed for the extraction of sucrose from sugarcane. At the end of the 1970s, it is noteworthy that European manufacturers adapted their equipment developed for extracting sugar beet into continuous diffusers to be used in the extraction of sugarcane bagasse.

To illustrate the expansion of the sugarcane sector in Brazil, between 1975 and 1985, more than four distilleries were installed with sucrose extraction technology through the diffusion process aimed at producing ethanol from sugarcane. The

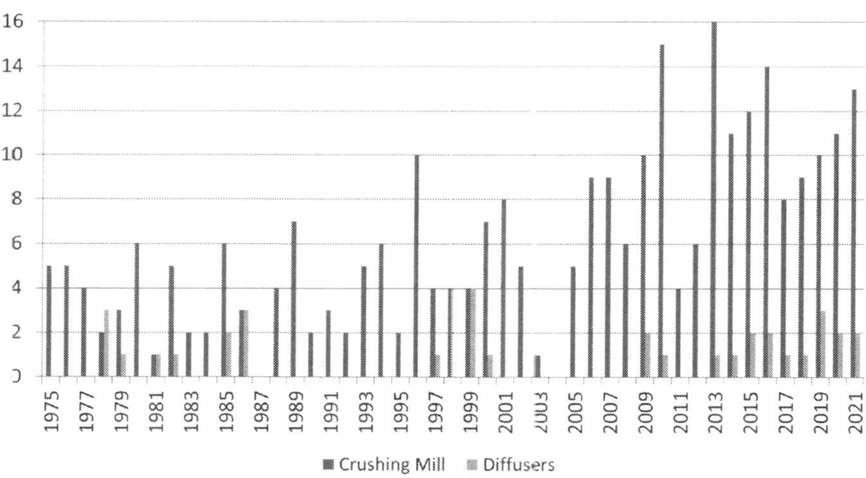

FIGURE 7.3 Search term results (crushing mill and diffuser) according to patent filing date (1975 to 2021).

Source: Authors.

diffusers, despite having been manufactured in Brazil, were licensed by European developers (Piacente, 2010).

At the end of the 1990s, the sugarcane sector in Brazil presented, mainly due to the increase in production, periods of investment. These investments in the sector contributed to the installation of new sugarcane processing plants aimed at producing ethanol in the country. Most of these new installations used diffusers to extract sucrose in more developed and efficient versions than those of the 1970s generation.

Figure 7.4 shows the documents researched for the term "crushing mill," according to the country of origin. It is observed that the largest number of documents for this search term was created in China (CN), representing 23.7% of the total documents. India (IN), the USA and Brazil (BR) also occupy a prominent position in this regard. There are 41 documents related to the acronym WO (Wipo), which refer to deposits made in more than one country at the same time.

Figure 7.5 shows the documents searched for the term "diffuser", according to the country of origin. In this case, the country with the largest number of documents is Germany (DE), with seven patent documents filed. In Brazil, two documents deposited as a priority country were verified.

Because a company has research centers distributed in several countries, it is difficult to specify which countries stand out in the technologies developed regarding the crushing mill and diffuser. The "priority country" information, where the patent document is first filed, highlights the need to protect the technology in the country of registration.

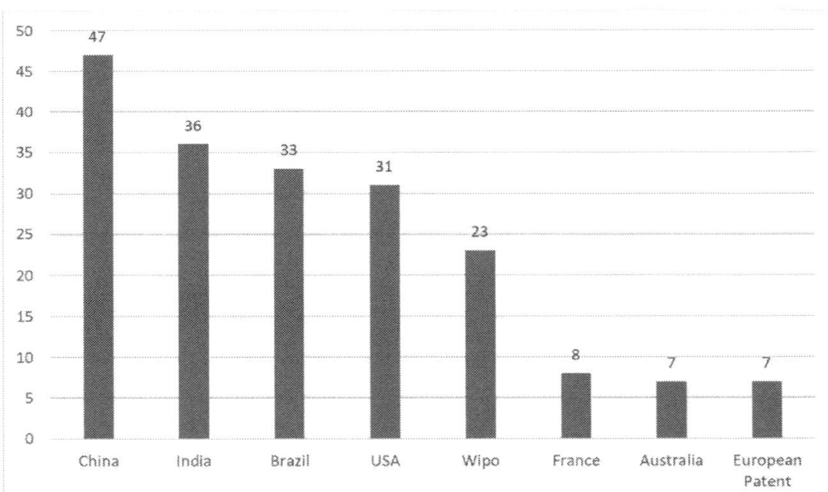

FIGURE 7.4 Main results obtained with the search term "crushing mill" by priority country from 1975 to 2021.

Source: Authors.

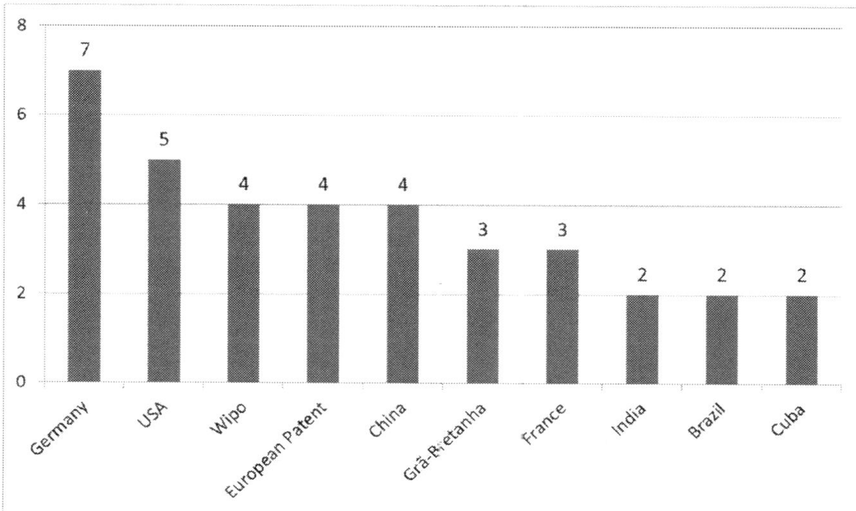

FIGURE 7.5 Main results obtained with the search term "diffuser" by priority country from 1975 to 2021.

Source: Authors.

The regulations of the international patent system allow a technology to be developed in a given country and registered in patent offices in another country, anywhere in the world. It is the firm's strategy to concentrate R&D centers in locations with greater comparative advantage and disseminate this developed technology in other countries where its application is required. Therefore, stating in which country the technological development of sucrose extraction systems, whether crushing mills or diffusers, is greatest is not an easy task. However, from the analysis of patent documents, it can be observed that countries with greater technological development in a given area have a greater probability of becoming the "priority country", that is, the place where it is carried out. This represents the first global patent filing for the technology, with this country serving as the primary jurisdiction for its protection. Furthermore, the patent document identifies the applicant, which is the individual or company that files the request to obtain rights over a creation that can be produced industrially. Identifying the profile of these depositors, who are the owners of the inventions, allows us to draw up a quantitative profile of the number of companies, educational institutions, R&D centers and individual agents who require the authorship of technologies developed in certain sectors of the economy.

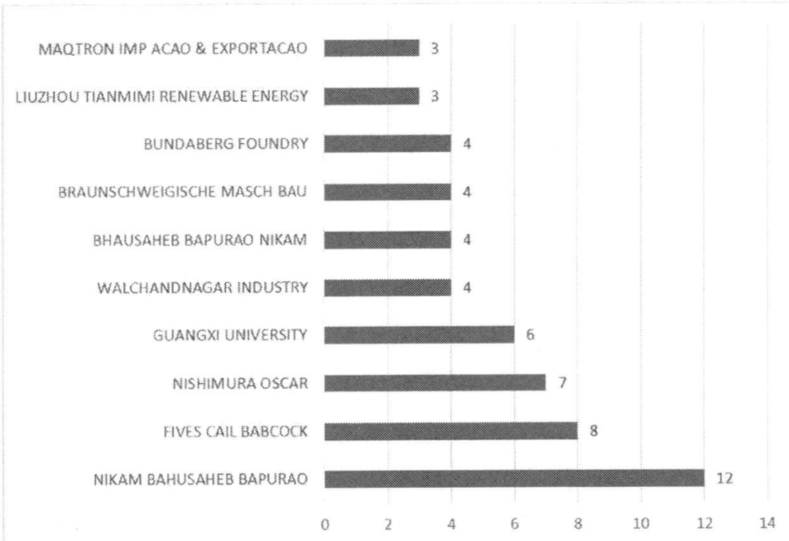

FIGURE 7.6 List of main applicants, actors and actors of patent applications with the search term "crushing mill" during the period 1975–2021.

Source: Authors.

Figure 7.6 shows the number of main patents requested in the search and the main applicants for these documents when the search term "crushing mill" was analyzed. It is noted that the main applicants come from (i) India: Bhausaheb Bapurao Nikan (16 documents), Fives Cail Babcok (8 documents), Walchandnagar Industries (4 documents); (ii) Brazil: Nishimura Oscar (7 documents), Maqtron Imp Ação & Exportação (3 documents); (iii) China: Guangxi University (6 documents), Liuzhou Tian Mimi Renewable Energy (3 documents); (iv) Germany: Braunschweigische Masch Bau (4 documents); (v) Australia: Bundaberg Foundry (4 documents).

Figure 7.7 presents the results of the search for the main patent documents requested between 1975 and 2021 that were related to the search term "diffuser" applied to the extraction of sucrose from sugarcane, separated by the author of the invention. It can be seen that the Austrian company BMA (Braunschweigische Masch Bau), a world leader in the development and production of machinery and equipment for the production of beet sugar, stands out as the main patent applicant with 24 documents requested, followed by the Czech Republic with 4 documents (Karel Cuker and Karel Comerce); from India with 2 documents (Spary Eng); Brazil and South Africa with 1 document each (Empral Desenvolvimento de Equipamentos and Sugar Mills Research Institute, respectively), among others.

Regarding the legal status of the patent documents searched using the search terms "crushing mill", it was observed that 44% of the documents analyzed are classified as "expired"; that is, the patent application was requested but was not validated by the requesting author within five years. Twenty-three percent of the documents were

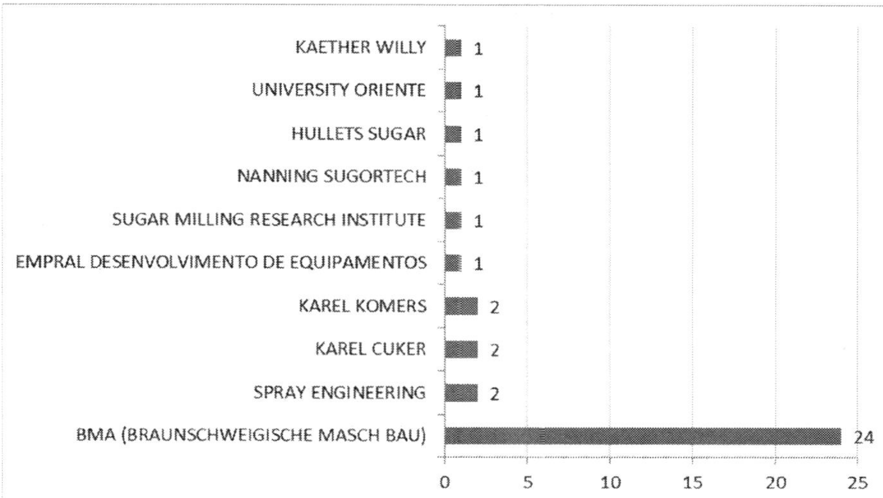

FIGURE 7.7 List of the main applicants of patent applications with the search term "diffuser" applied to the extraction of sucrose from sugarcane between 1975 and 2021.

Source: Authors.

identified as "pending", indicating that there is some type of document non-conformity or lack of payment that allows their registration to continue. Fifteen percent of the documents analyzed were identified as "expired"; that is they became public domain. The rest were classified as follows: 3% granted, those whose request was granted and the holders received payment for the license; and 6% were considered "revoked"—no payment for the patent was made by the applicant. Regarding the document analyzed using the search term "diffuser", 43% of the documents searched were considered "expired"; 25% granted"; 16% "expired"; 8% presented as "pending"; and 9% "revoked".

7.5 CONCLUSION AND FUTURE OUTLOOK

The objective of this chapter was to identify the technological evolution of two processes used in the extraction of sucrose from sugarcane: crushing mill and diffuser. To this end, a search was carried out in patent databases from the main industrial protection offices in the world, using a specific search tool for this purpose, the Questel Orbit software.

Currently, research of this type has gained importance and relevance in the scientific scenario around the world. It is an alternative to quantify and diagnose the state of the art and the technological trajectory of a certain applied technology and important economic sectors. This research identified the researched patent banks which are the main applicants and authors of the main technologies related to the search terms researched. It pointed out the "priority country" in which this technology was initially protected in the form of a patent application, among other relevant information.

From an applied point of view, this research allows the generation of subsidies so that scientific, technological and R&D policies can be developed more accurately in an important economic sector around the world, such as sugar and bioethanol producers.

The data obtained in this research made it possible to highlight important aspects about the development of scientific and technological research in the area in question. It was noted, through patents registered as a "priority country", that China, India and Brazil are the main individual developers of technologies that use the "crushing mill" process to extract sucrose from sugarcane. In this context, China was the leader with 68 patent documents filed between 1975 and 2021, followed by India with 42 and Brazil with 35.

As for the data obtained from patent databases, when the search term was related to the "diffusion" process, Germany presented itself, in the period analyzed, as the "priority country" with the largest number of documents deposited. This fact is justified; the technology for extracting sucrose from sugarcane using "diffusion" originated in an adapted way from the production of beet sugar, which uses only "diffusion" devices in this process. Thus, it can be seen that BMA, the main developer of technology focused on the production of equipment for processing beetroot, has successfully adapted its innovations for the processing of sugarcane.

In Brazil, the National Alcohol Program (ProÁlcool), created in 1975, consisted of an initiative by the Brazilian government to intensify the production of fuel alcohol (ethanol) to replace gasoline. This program offered several tax incentives and loans to sugarcane producers, technology developers for the sector and the automobile industry, for the development of alcohol-powered cars. Programs of this type encouraged the capital goods industry linked to sugarcane processing, enabling the development of new technologies and their industrial protection through patents.

REFERENCES

Breward, C., Hocking, G., Ockendon, H., Please, C., & Schwendeman, D. (2012). *Modelling the extraction of sugar from sugar cane in a diffuser*. Johannesburg: Proc. MISG South Africa, University of the Witwatersrand, pp. 31–57.

Castro, T., Cavallaro, R., Bucek, E., & Begnini, M. (2018). Comparação da eficiência de extração da sacarose da cana de açúcar entre a moenda e o difusor. In: *II encontro de desenvolvimento de processos agroindustriais*. Uberlandia: UNIUBI.

Cavalcanti, D. A. B. (2005). Moenda X Difusor: Qual o melhor processo de extração? *Revista Opiniões*, outubro/dezembro de.

FAO—Food and Agriculture Organization. (2022). *FAOSTAT*. Available at: http://faostat.fao.org/.

Macedo, M. F. G. (2000). *Patentes, pesquisa & desenvolvimento: Um manual de propriedade industrial*. Rio de Janeiro: Fiocruz.

Piacente, F. J. (2010). *Inovação e trajetórias tecnológicas: O caso dos dois sistemas para extração de sacarose no Brasil*. Campinas: IE/Unicamp. Tese do Doutorado.

Piacente, F. J., & Silva, V. C. (2015). Setor sucroenergético Brasileiro: Um estudo de prospecção tecnológica e inovação entre 1975 e 1985. *Revista Espacios*, v. 36, n. 23, pág. 12.

Piacente, F. J., Silva, V. C., & Armas, O. F. (2021). Evolução da produção agrícola canavieira na região noroeste do estado de São Paulo: Fases de expansão e crise no setor e seus impactos no uso da terra entre 2000 e 2013. *Revista de Economia e Sociologia Rural*, v. 60, n. 1, pp. 1–16.

Piacente, F. J., Silva, V. C., & Biaggi, D. E. (2015). Produção de etanol de segunda geração a partir da cana-de-açúcar: Estudo de prospecção de patentes. *Revista Espacios*, v. 36, n. 23, pág. 16.

Piacente, F. J., Silva, V. C., & Biaggi, D. E. (2016). Sugar cane juice extraction innovation: Technological prospecting study about diffuser and crush process. In: *POMS 27th Annual Conference*, Orlando.

Rein, P. (2013). *Engenharia do açúcar de cana*. Edição em Português. Editora Bartens, p. 608.

Silva, E. F., Borges, E. S. S., Rocha Porto, P. C., & Peralta, P. P. (2021). *Patente: Da importância e sua proteção: Patente de invenção e modelo de utilidade*. Rio de Janeiro: INPI.

Stupiello, J. P. (1987). A cana de açúcar como matéria prima. In: *Cana de açúcar cultivo e utilização*. Campinas, SP: Fundação Cargil, Volume 2.

USDA. (2022, April 15). United States Department of Agriculture. *World Agricultural Supply and Demand Estimates and National Agricultural Statistics Service: Sugar Annual*.

8 Comparative Analysis of Recycled HDPE Tiles with Graphene against Solar Panels

A Study from the Perspective of Frugal Innovation and Technoscience as an Instrument for Technological and Social Development

Sergio Assis Machado, Rosinei Batista Ribeiro, Adison da Silva Mello, Francisco Del Moral Hernandez, Caio Fernando da Silva, and Ramón Quiza Sardiñas

8.1 INTRODUCTION

In the context of the search for sustainability and the goals to be met by nations, solar panels and graphene roof tiles are elements that generate clean and sustainable energy. Recycled HDPE roof tiles with graphene have emerged as another alternative to conventional roof tiles and solar panels as an element in energy generation. Due to the properties of graphene, the roof tiles have a longer lifespan (around 80 years), energy generation capacity, and lower cost than solar panels, as well as covering capacity, and the fact that they produce clean and environmentally friendly products, due to the plastic being of recycled origin, a sustainable product.

Assuming these characteristics, the research seeks to assess whether the applications of recycled HDPE roof tiles with graphene can be classified as a frugal innovation product, a way of doing more with less and optimizing resources for the benefit

 DOI: 10.1201/9781003502647-10

of the population, comparing them with solar panels, and whether this technology can be classified as a technoscience concept.

In this scenario of frugal appeal, roof tiles can have the connotation of a social innovation product with techno-scientific characteristics, where a product, good, or service, under the action of a social actor in the production process, modifies it in a quantitative or qualitative way, seeking to meet social needs within a given socio-economic context.

Within this context, the chapter explores the capacity of artifacts in the frugal innovation scenario based on technoscience with the purpose of producing clean energy accessible to a larger portion of the population by comparing the energy efficiency of recycled HDPE roof tiles with graphene and solar panels.

8.2 LITERATURE REVIEW

Frugal innovation initiatives are usually attributed to countries with emerging economies where there is a large part of the population at the bottom of the pyramid, with social needs to be met. However, this approach has become relevant in industrialized nations, affecting the competitiveness of local industries in the long term and abroad (Radjou & Prabhu, 2015).

Frugal innovation aims to achieve more with less, developing new products, services, and processes that meet the essential needs of consumers but with reduced resources (Suny et al., 2011). Compared to Jugaad innovation, which means improvisation in Hinduism, frugal innovation requires agile adaptation before concrete guarantees of a solution. Radjou et al. (2012), Cassiolato et al. (2020) and Zeschky et al. (2014) highlight three fundamental principles: (1) the importance of doing more with less, finding opportunities in difficulty; (2) the defense of the flexible mind of the researcher, with less commitment to pre-established standards; (3) the emphasis on simplicity, avoiding luxury or sophistication.

It can also be defined as a solution to scarce resources, developed under resources and constraints, to produce affordable and acceptable goods to serve customers who cannot afford the conventional equivalent (Oliveira, 2004; Zeschky et al., 2014).

Frugal innovation promotes sustainability more than conventional products, and, in addition, sustainable business models have competitive advantages by capturing and creating value propositions where solutions are maximized (Livari et al., 2009).

Thus, frugality seeks to create basic products and services for a large number of people, following the idea of "doing more with less" and providing greater value and benefits by using fewer resources (Tripathi et al., 2017). In short, frugal innovation seeks simple and efficient solutions to complex problems, being especially relevant in contexts of scarce technological, infrastructural, and financial resources, opting to extract the maximum with minimal resources and adapting technologies.

Reasons to invest in frugal innovation:

- Slow growth in developed economies will increase demand for frugal innovations.
- Environmental restrictions have increased demand for frugal models of production and consumption.

- Aging societies have demanded new approaches to health and social care requirements.
- Understand that the fastest-growing markets are in developing economies, where frugal services and demand are high.

According to Koerich et al. (2019), a more sustainable economy offers consumers the opportunity to meet their needs without compromising the future of generations to come, just as frugal innovation does.

8.3 SOCIAL INNOVATION

The term "innovation with social impact" includes innovative products, processes, and services that address social challenges. The term "social innovation", as it is known today, gained importance in academia with the work of Taylor (Sun et al., 2011).

A broad approach can include innovations developed by large companies, protected by industrial property rights, and whose generation is aimed at profit, as long as they have a positive impact on social demands or needs. Cassiolato et al. (2020) cite as an example both high-tech solutions aimed at public health and low-tech solutions, such as the diffusion of cisterns in the Brazilian semi-arid region. Social innovation is based on social theory (which deals with the interaction between various agents) and is therefore located at the interfaces between government, business, and civil society, which have distinctions and are largely protected from each other, in their respective logics of action and regulatory mechanisms, as well as associated problems and limited problem-solving capacities (Livari et al., 2009).

"Innovations with a social impact" include a specific segment, generically called "social innovation or technology", in which groups not only benefit from such technologies but also collaborate directly in their development. The construction of social reality is based on relationships of conflict and cooperation between actors, who have different ways of explaining reality, given that they may be in different situations (Radjou & Prabhu, 2014).

Social innovation is concerned with reducing poverty or meeting social needs, improving living conditions, and seeking the well-being of the community, which, according to the analysis, presents the following barriers (Feitosa et al., 2022) (Figure 8.1).

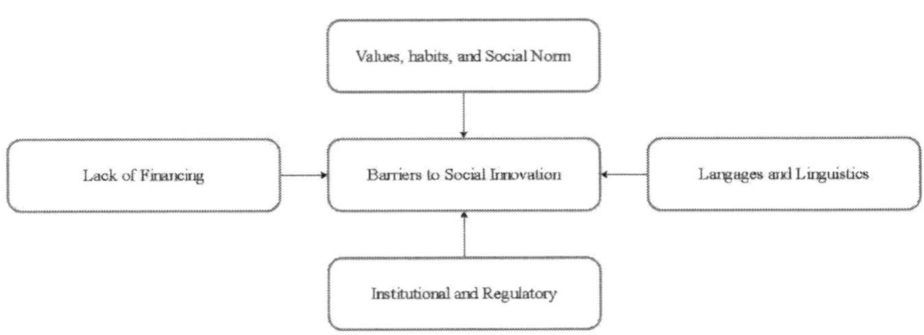

FIGURE 8.1 Barriers to social innovation (Feitosa et al., 2022).

8.4 SUSTAINABILITY AND SUSTAINABLE DEVELOPMENT

Sustainability is a concept that is strongly related to three pillars: social, environmental, and economic; together, they ensure the conditions for the survival of the planet and are closely related to the concept of "sustainable development", which is defined as meeting the needs of present generations without compromising the ability of future generations to meet their needs (Teixeira, 2023). This concept was first defined in the 1987 Brundtland Report.

With this in mind, the United Nations (UN) established the 17 Sustainable Development Goals (SDGs) in 2015, and they form a global agenda for the construction and implementation of public policies aimed at guiding humanity until 2030.

The agenda includes an international action plan to achieve the 17 SDGs, broken down into 169 targets, which address various fundamental issues for human development, from five perspectives: people, planet, prosperity, partnership, and peace.

This mode of development has been defended by the UN since the New York Summit in 2015, when the then Secretary-General, Ban Ki-moon, said: "The agreement encompasses a universal, transformative and integrated agenda (. . .) This is the People's Agenda, an action plan to end poverty in all its dimensions, irreversibly, everywhere, leaving no one behind" (ONU, 2015). What has been observed, and which is consistent with the reality analyzed by Amartya Sen, is that even in very economically rich countries there are groups with lower expectations and quality of life, for whom economic security is not a reality, and in countries with democratic problems, the situation of inequality is even greater (Batista, 2019).

Sustainable development must promote equal rights and opportunities for all and lead to greater balance and fairness in the distribution of wealth and access to the resources and services essential for a dignified life, such as education, health, food and housing, and human rights, to promote greater equity for the most disadvantaged groups, enabling and creating the conditions for them to thrive and grow independently.

8.5 GRAPHENE

Graphene is a flat layer of carbon atoms arranged in a hexagonal crystalline structure, only one atom thick, and is the first "2D material" available for scientific and technological use (Bellucci et al., 2021). Although the term "graphene" is also used to describe materials with multilayers, structural defects, or functionalization, which differ from the theoretical definition, these still have highly distinct and specific parameters.

Essentially, graphene consists of a sheet of carbon atoms arranged in a hexagonal structure, similar to honeycombs, and is an allotrope of carbon (Araujo, 2022). Its properties include greater strength than steel, greater conductivity than copper, greater electron mobility than silicon, and the ability to be transparent (Marin, 2021).

Carbon–carbon bonds are the strongest bonds found in nature, where each carbon joins three others in the structure.

Therefore, the hybridization of the atom is *sp*2, which corresponds to two single bonds and one double bond (Figures 8.2 and 8.3).

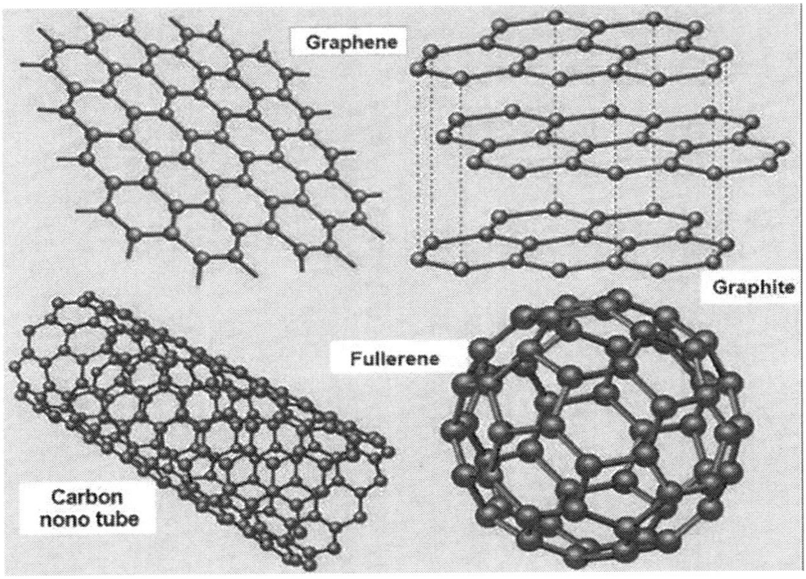

FIGURE 8.2 Schematic structure of the graphene sheet (Batista, 2019).

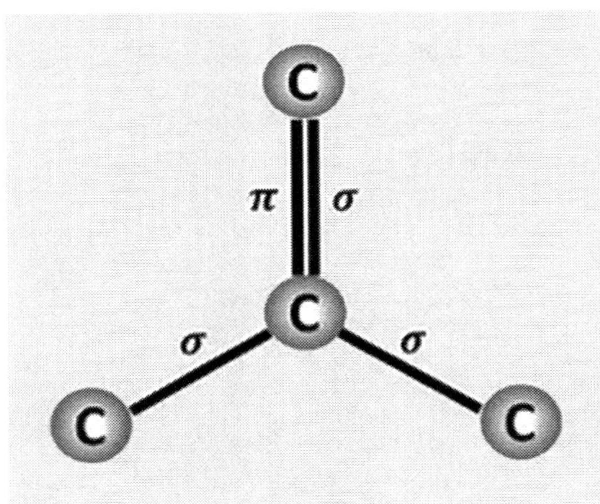

FIGURE 8.3 sp2 hybridization of carbon in graphene (Batista, 2019).

The addition of graphitic nanofillers to the polymer matrix offers the opportunity to improve polyolefins such as HDPE. Generally, the insertion of small amounts of the nanofiller into the polymer matrix is enough to improve the physical, mechanical, and thermal properties of polyolefin for advanced engineering applications (Sun et al., 2011).

8.6 SOCIAL TECHNOLOGY

Social technology (ST) is understood as "a set of transformative techniques, methodologies, developed and/or applied in interaction with the population and appropriated by them, which represent solutions for social inclusion and improvement of living conditions" (Caderno de Debate—Social Technology in Brazil).

The concept of ST establishes four dimensions:

1. Knowledge, science, and technology ST starts with social problems; ST is organized and systematized; ST introduces or generates innovation in communities.
2. Participation, citizenship, and democracy ST emphasizes citizenship and democratic participation; ST adopts participatory methodology in work processes; ST promotes its dissemination and reapplication.
3. Education ST carries out an entire pedagogical process; ST develops in a dialogue between popular and scientific knowledge; the communities, which gain autonomy, appropriate ST.
4. Social relevance ST is effective in solving social problems; ST is environmentally sustainable; ST brings about social transformation. Until 2015, social technologies contributed, in a participatory and democratic way, to the UN Millennium Development Goals. Now, with the UN's 2030 Agenda and its 17 SDGs, social technologies are important instruments for building a fairer, more resilient, and sustainable world (Ministério da Ciência e Tecnologia e Inovação, 2025).

8.6.1 Technoscience

Technoscience is the conscious manipulation of work processes by a social agent, resulting in the alteration of the final product according to the socio-economic context, social norms, and production environment. Differentiating itself from simple innovation and technology, technoscience stands out as a flexible and adaptive approach to managing diverse knowledge, promoting the appropriation of results by social agents in search of changes in production and consumption systems. This approach values collective ownership of the means of production and self-management, aiming to meet community needs through solidarity economy networks (Dagnino & Gomes, 2020).

Idealized as a programmatic alternative to terms such as innovation (social, responsible, open, etc.) and technology (social, sustainable, etc.), it can be understood as a way (original, open, mutating, and adaptive) of acting on knowledge of any kind (scientific, empirical, social, and technological), it can be understood as a (original, open, mutating and adaptive) way of organizing (often through socio-technical adaptation of capitalist technoscience) knowledge of any nature (scientific, empirical, technological, religious, ancestral . . .) and origin (academic, business, native peoples, popular movements, excluded . . .) by social actors with the aim of appropriating a material result derived from changes in the process of production and consumption of goods and services in solidarity economy networks, respecting their values and interests (collective ownership of the means of production, self-management, etc.), oriented primarily toward satisfying collective needs (Dagnino & Gomes, 2020).

In generic terms, technoscience is the cognitive result of an actor's action on a work process that they control and which, depending on the characteristics of the socio-economic context, the social agreement, and the productive space in which they operate, allows a modification to the product generated that can be appropriated, according to their interests. Or, more simply, technoscience is the cognitive consequence of the action of a social actor on a work process that they control, which allows a modification (qualitative or quantitative) to the product generated (in the generic sense of output) that can be appropriated according to their interests (Dagnino, 2019).

According to Oliveira (2004), the conflict between science and technology is also responsible for the characteristic problems of modern society. He argues that, in the current context of liberal hegemony, technoscience shares responsibility for the problems that plague humanity, such as the persistence of hunger, poverty, violence, and social inequalities. It also contributes to environmental degradation, including the depletion of natural resources.

8.6.2 Solar Panel Controversies

Solar panels are manufactured to generate electricity and have a social, economic, and environmental appeal, as they are an artifact that can be transformed, provide renewable and environmentally friendly energy, and have a social and democratic appeal.

However, when looked at from another angle, solar panels contain elements that can lead to divergent opinions about them.

According to Latour, things are more than tools, a backdrop or a stage on which human social actors play the main roles, proposing a perspective that advocates equal status and attention for human and non-human actors.

Manufacturing solar panels requires the use of a lot of energy and materials—water, metals, plastics, glass, and other parts and components that produce social and environmental impacts (Anselmo, 2019).

Fthenakis takes an important approach to the environmental impacts produced during the life cycle of photovoltaic panels (Fthenakis, 2018).

Mulvaney (2019) argues that the manufacture of solar panels involves the use of toxic chemicals and production processes that have negative environmental impacts.

However, according to Santi et al. (2021), photovoltaic solar energy plays a crucial role in promoting sustainability, since it is a clean and renewable energy source that contributes significantly to reducing dependence on fossil energy resources and mitigating negative impacts on the environment.

Ghizoni (2016) states that, through recycling, waste can be returned to the production chain as raw material, generating savings by reducing costs and increasing the competitiveness of new products, thus avoiding further exploitation of sources that are still conserved.

Recycling worn-out modules may not be seen as an immediate problem in the development of solar energy.

However, the rapid growth of solar energy will result in waste disposal problems in the near future, as the end of life of photovoltaic panels will generate a significant amount of waste.

According to Tolmasquim (2004), the photovoltaic system generally has the following negative environmental impacts: emissions and other impacts associated with the production of energy required for the manufacturing, transportation, installation, operation, maintenance, and decommissioning processes of the systems; emissions of toxic products during the raw material process for the production of modules and peripheral components, such as acids and carcinogenic products, as well as CO_2, SO_2, NO_x, and particulates; occupation of area for project implementation and possible loss of habitat (critical only in special areas)—however, photovoltaic systems can make use of existing areas and structures such as roofs and façades; occupation of area for project implementation and possible loss of habitat (critical only in special areas)—however, photovoltaic systems can make use of existing areas and structures such as roofs and façades; visual impacts, which can be minimized by choosing non-sensitive areas; risks associated with the toxic materials used in photovoltaic modules (arsenic, gallium, and cadmium) and other components.

According to the Brundtland Report, sustainability is the concept according to which human activities should be guided by respect for the environment and the rational use of natural resources (World Commission on Environment and Development, 1987).

However, the extraction of materials such as silicon and silver, used in the production of solar cells and solar panels, may be unsustainable in the long term, considering the finiteness of these natural resources (Anselmo, 2019).

Another position that causes debate about the life cycle of panels refers to the generation of electricity without emissions during their operation, the so-called carbon footprint, which is the total greenhouse gas emissions associated with their manufacture, which can be considerable, generating controversy regarding their effectiveness in terms of net reduction of emissions throughout the life cycle.

Another point that, although it can be identified as a metaphor, is controversial is the fact that, according to the definition of Farm in the Michaelis Dictionary (2023), it refers to "large rural property used for farming or raising livestock".

However, another definition has been created for farms that are actually power stations; "solar farms", which, according to Amaral et al., are those installed on the ground, with their support structures (foundations) and fixings (fittings) installed directly on an earthen area and dedicated exclusively to solar energy generation.

This point is criticized because centralized solar power plants, where solar panels are installed over large areas, often occupying several hectares of land, cause environmental impact on the local biotic environment and possible damage to the ecosystem. The panels can also cause damage to the vegetation due to earthworks, shading of the soil, possible risks to local animals due to accidents, and compromised vegetation that serves as food.

For these reasons, scientists such as Mulvaney (2019) in the United States is critical of this type of solar arrangement (Mulvaney, 2019).

According to the REN21 report (2020), the expansion of solar photovoltaic energy not only contributes to environmental sustainability but also promotes economic and social benefits, including job creation and the reduction of regional inequalities. However, implementing solar energy on a large scale can benefit wealthier communities more, perpetuating socio-economic disparities, since the initial installation can require substantial investments.

Rather than serving as a social benefit for the neediest sections of the population, they favor the higher social classes by the way they are used at the moment, to the detriment of the lower ones who really need them.

Although the cost of solar panels has fallen, the initial investment for installation can still be prohibitive for many families and companies, becoming an economic barrier.

For science, according to Latour (2012), it analyzes the actors involved, and as it is not linear, it transcends the controversies and clarifies the meaning of this network of human and non-human actors, and the controversies aligned between the lines deconstruct the meanings of artifacts and answer many of the questions about the absolute values of science.

Latour also points out that technologies carry values, interests, and powers that shape society in a complex way and that it is important to consider the political, social, and technological dimensions rather than just treating them as objective tools. Therefore, it is worth noting that, although there are controversial opinions, solar panels have benefits, and for a truly sustainable transition to renewable energy generation, it is necessary to critically analyze existing positions and adopt ethical practices in the production, disposal, and life cycle of solar panels to mitigate negative impacts.

8.7 METHODOLOGY

The methodology to be applied in the research is design science research (DSR): according to Simon (1996), design science involves a pragmatic research paradigm that seeks to create innovative artifacts to solve real-world problems. DSR includes this focus on the artifact, as well as giving priority to the relevance of its application. DSR seeks to solve problems rather than explain an existing reality or help to understand it (Iivari & Venable, 2009).

Table 8.1 presents a comparison between Traditional Research and research conducted using DSR.

According to Dresch et al. (2020), DSR "is the method that underpins and operationalizes the conduct of research when the objective to be achieved is an artefact or a prescription".

By development, we are not referring solely and exclusively to product development. DSR can serve this purpose but has a broader objective: to generate knowledge that is applicable and useful for solving problems, improving existing systems, and creating new solutions and/or artifacts.

The artifact is any human creation that has not been contemplated by its origin in nature; in other words, it is a product of human knowledge and technique.

The fundamental characteristic of research that uses DSR is its orientation toward solving specific problems, not always looking for an optimum solution, but a satisfactory solution that satisfies the problem situation (Dresch et al., 2020).

TABLE 8.1
Comparison between Design Science Research and Traditional Research

Characteristics	Design Science Research	Traditional Research
Objectives	Developing artifacts that enable satisfactory solutions to practical problems Prescribing and designing	Explore, describe, and explain
Main activities	Raise awareness—suggest—develop evaluate—conclude	Plan action—collect data—analyze data and plan action
Results	Artifacts (constructs, models, methods, instantiations)	Constructs, hypotheses, descriptions, explanations, actions
Type of knowledge	How things should be	How things are or how they behave
Role of the researcher	Artifact builder and evaluator	Multiple, depending on the type of action research
Empirical basis	Not required	Not required
Researcher researched collaboration	Not required	Obligatory
Implementation	Not required	Obligatory
Evaluation of results	Applications, simulations, experiments	Confrontation with theory
The approach	Qualitative and/or quantitative	Qualitative

Source: Adapted from Dresch et al. (2020).

Two factors are fundamental to conducting successful DSR research: rigor, which is essential to contribute to increasing the knowledge base in a given area and to be considered valid and reliable; and relevance, because for the professionals involved it is essential to use the results of the research and the knowledge generated to solve practical problems (Dresch et al., 2020).The guidelines for conducting research using the DSR methodology are presented in Table 8.2, adapted to the artifacts.

The DSR methodology can support the development and construction of artifacts based on the needs observed, as well as the problems that interest the researcher, helping to strengthen the knowledge base, providing the ideal environments for testing new theories and developing new artifacts (Dresch et al., 2020).

A comparison was made between the characteristics of a recycled HDPE roof tile with graphene, Figure 8.5 and a 150 w solar panel, Figure 8.6, which were studied in relation to their energy efficiency. The daily and monthly energy generation capacities were calculated and applied to an energy consumption range of 30 kWh, which can be extrapolated to higher consumptions.

An exploratory survey of theory and ideas on frugal innovation, ST, and technoscience was carried out to determine whether the artifact satisfies the research objective (Figure 8.4).

The classification of the research used in the work was of an applied nature, with exploratory objective, qualitative, and quantitative approaches, and the DSR method was adopted.

TABLE 8.2
Guidelines for Conducting Original Design Science Research

Guidelines	Concept
Designer as artifact	Research carried out using the DSR method must produce viable artifacts in the form of a construct, model, method, or instantiation.
Relevance of the problem	DSR's goal is to develop solutions to solve important and relevant problems for organizations.
Designer evaluation	The usefulness, quality, and effectiveness of the artifact must be rigorously demonstrated through well-executed evaluation methods.
Research contributions	Research conducted using the DSR method should provide clear and verifiable contributions in the specific areas of the artifacts developed and present a clear grounding in design fundamentals and/or design methodologies.
Rigorous research	The research must be based on an application of methods, in both the construction and evaluation of the artifacts.
Design as a research process	The search for an effective artifact requires the use of means that are available to achieve the desired ends, while satisfying the laws governing the environment in which the problem is being studied.
Research communication	Research conducted using the DSR method should be presented to both a more technology-oriented audience and a more management-oriented one.

Source: Adapted from Hevner et al. (2004) and Dresch et al. (2020).

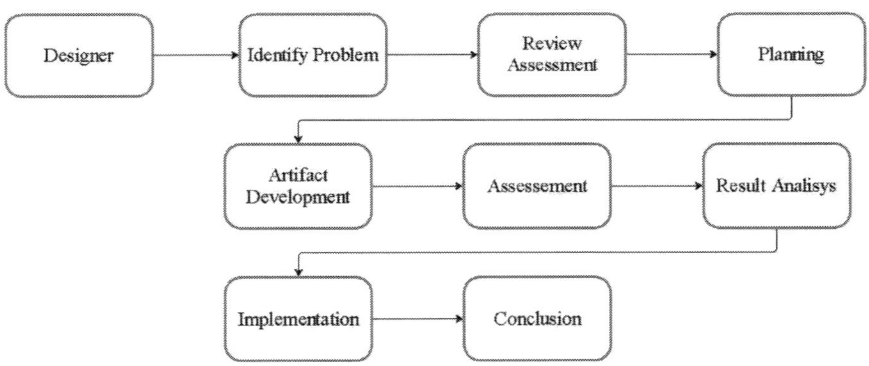

FIGURE 8.4 Guidelines for conducting the source design science research.

Source: Authors.

The development of the work was based on the application of DSR to identify the research problem, which was to compare the energy efficiency of HDPE roof tiles with graphene inserts and solar panels, using the characteristics of the elements presented in Table 8.3.

Two equations (8.1), (8.2), (8.3), and (8.4) were developed to calculate energy efficiency based on a consumption of 30 kWh; this consumption satisfies the consumer in the social tariff range and the characteristics of the artifacts.

FIGURE 8.5 HDPE tiles with graphene.

FIGURE 8.6 Solar panel.

Source: Made in China (2022).

TABLE 8.3
Characteristics of the HDPE Roofing Sheet with Graphene and the Solar Panel Used

Construction Elements Products	HDPE Tiles with Graphene Pallets	Solar Panels 150 w
Weight	7 kg	11.5 kg
Measurements	1.10 x 19.90 m	1.49 × 0.66 m
Energy generation	7.5 kWh/month	15.18 kWh/month
Lifespan	80 years	30 years
Cost	U$29.46	U$55.97

Source: Authors.

TABLE 8.4
Result Amount, Energy, and Costs

	HDPE Tiles with Graphene Pallets	Solar Panels 150 w
Parts	4	2
Total generation (Get)	30 kWh	30.36 kWh
Cost	~U$117.83	~U$111.94

Source: Authors.

It was considered:

where: Gmt : monthly generation tiles $n^i t$: number of parts tiles
Gdt : daily generation tiles $n^i p$: number parts panel
Gmp : monthly panel generation Get : total generation
Gdp : daily panel generation

1. $Gdt = \frac{Gmt}{30\,day}$
2. $Gdp = \frac{Gmp}{30\,day}$
3. $Gmt * n^i t = Get$
4. $Gmp * n^i p = Get$

8.8 RESULT

The results presented were obtained considering the consumption of 30 KWh/month by a low-income household in Brazil benefiting from the social tariff, according to the Ministry of Mines and Energy (2022), which receives the benefit of paying 65% of the basic tariff and may be the biggest beneficiaries of the use of the devices.

The calculations are presented below, and the results are shown in Table 8.4.

HDPE tiles:

$$Gmt = 7{,}5KWh/month$$

$$Gdt = \frac{Gmt}{30\,days} = \frac{250\,Wh}{day}$$

$$Gmt * 4 = \frac{30\,KWh}{month}$$

150 W solar panel:

$$Gmp = 15{,}18\,KWh/month$$

$$Gdp = \frac{Gmp}{30\,day} = \frac{506\,Wh}{day}$$

$$Gmp * 2 = \frac{30{,}36\,KWh}{month}$$

8.9 RESULTS AND DISCUSSION

The costs of insulated artifacts are equivalent for approximately the same energy generation, but if we consider that roof tiles are made from recycled HDPE and that their cost, according to Telite (Ciclovivo, 2021; Marin, 2021), is 40% lower than solar panels. The cost of installing a solar panel is higher than that of roof tiles, which are lighter, have more covering capacity, and require less structure to support them. According to the CUB (basic unit cost) index established by Brazilian National Law No. 4591/64, calculated by the Civil Construction Union in each region, the roof covering structure represents between 3.0% and 6.5% of the cost of a construction project (Sienge, 2025).

Consumers covered by the social tariff have their energy needs met with two graphene tiles, with a lower generation cost and consequently a lower installation cost, which would benefit a larger portion of the population through inclusion in energy generation and gaining benefits.

REFERENCES

Anselmo, A. Reciclagem ou destinação final dos Painéis Fotovoltaicos aplicados em geração de energia ao final do ciclo de vida. Universidade Federal de Minas Gerais, 2019.

Araujo, C. A. L. A. Grafeno: Composição, Propriedades, Aplicabilidade e Perpectivas. Revista brasileira de Processos Químicos. dez. 2022.

Batista, C. "Grafeno: O que é, aplicações, estrutura e propriedades." (2019). Accessed in: 22 ago. 2025: https://www.todamateria.com.br/grafeno.

Bellucci, F., Vasquez, H., & Conti, J. Panorama Tecnológico Grafeno. FINEP/MCTI, Rio de Janeiro, 2021.
Cassiolato, J., Soares, M. C. C., & Lastres, H. Innovation in unequal societies: How can it contribute to improve equality? In: Seminario Internacional, 2020.
Ciclovivo. (2021). Telhas Solares Geram Energia com Grafeno. Accessed in: 23 ago. 2025: https://ciclovivo.com.br/planeta/energia/telhas-solares-geram-energia-com-grafeno/.
Dagnino, R. Tecnociência Solidária: um manual estratégico. 1°edição, Marilia, 2019.
Dagnino, R., & Gomes, E. Sistema de Inovação Social para Prefeituras. In: Conferência Nacional de Ciência e Tecnologia para Inovação. Anais São Paulo, 2020.
Dicionário Michaelis. Editora Melhoramentos LTDA. 2023.
Dresch, A., Lacerda, D. P., & Antunes, J. A. V. Design science research: Método de pesquisa para avanço da ciência e tecnologia. Porto Alegre: Bookman, 2020.
Feitosa, F. J. S., Sano, H., Ramos, A. S. M. (2022). Barreiras e indutores da inovação social: uma revisão sistemática da literatura. Studies in Social Sciences Review, Curitiba, 3(3), 768–783.
Fthenakis, V. M. E. L. P. A. Electricity from sunlight: Photovoltaic systems integration and sustainability. 2nd ed. [S.l.]: Hoboken, NJ: John Wiley & Sons, 2018.
Ghizoni, J. P. Sistemas fotovoltaicos: estudo sobre reciclagem e logística reversa para o Brasil. Universidade Federal de Santa Catarina, 2016.
Hevner, A. R., March, S. T., Park, J., & Ram, S. (2004). Design science in information systems research. MIS Quarterly, 28(1), 75–105.
Iivari, J., & Venable, J. Action research and design science research—seemingly similar but decisively dissimilar. In: European Conference on Information Systems, 17. Verona, 2009.
Koerich, G. V., & Cancellier, É. L. P. D. L. (2019). Inovação Frugal: Origens, evolução e perspectivas futuras. Cadernos Ebape. Br, 17(4), 1079–1093. https://doi.org/10.1590/1.
Latour, B. Reagregando o social: Uma introdução à Teoria do Ator-Rede. Salvador: EDUFBA, 2012.
Made in China. (2022). Figure Solar Panel. Accessed in: 23 ago. 2025: https://jh-888.en.made-in-china.com/product/WvyJDNgoMBRj/China-Solar-Wholesale-Factory-Price-Dual-Axis-Sun-Tracking-Tracker-System.html.
Marin, J. Telha solar com grafeno é aposta da Telite para gerar energia. Temundo, 2021. Accessed in: 22 ago. 2025: https://www.tecmundo.com.br/ciencia/219944-telha-solar-grafeno-aposta-telite-gerar-energia.htm.
Ministério da Ciência e Tecnologia e Inovação. (2025). Accessed in: 23 ago. 2025: https://antigo.mctic.gov.br/mctic/opencms/ciencia/politica_nacional/_social/Tecnologia_Social.html.
Ministry of Mines and Energy. (2022). Tarifa Social. Accessed in: 23 ago. 2025: https://www.gov.br/aneel/pt-br/assuntos/tarifas/tarifa-social.
Mulvaney, D. Solar power: Innovation, sustainability, and environmental justice. Okland: University of Califórnia Press, Green Energy, SAGE Publications, April 2019.
Oliveira, M. B. Desmercantilizar a Tecnociência. In: Santos, B. S. (org.). Um Discurso sobre a ciência revisitado. São Paulo: Cortez, 2004.
ONU. (2015). Transformando nosso mundo: A agenda 2030 para o Desenvolvimento Sustentável. Accessed in: 22 ago. 2025: https://brasil.un.org/pt-br/91863-agenda-2030-para-o-desenvolvimento-sustent%C3%A1vel.
Radjou, N., & Prabhu, J. Frugal innovation: How to do more with less. London: Profile Books, 2014.
Radjou, N., & Prabhu, J. Frugal innovation: How to do more with less. New York: Public Affairs, 2015.
Radjou, N., Prabhu, J., & Ahuja, S. Jugaad innovation: Think frugal, be flexible, generate breakthrough growth. San Francisco: Jossey-Bass, 2012.

REN21. (2020). *Renewables 2020 – Global status report.* Paris: REN21 Secretariat. Accessed in: 23 ago. 2025: https://www.ren21.net/reports/global-status-report/.

Santi, J. R., Luchesi, L. F., & Rosa, V. L. (2021, 10 de dezembro). Energia solar limpa, renovável e inesgotável. Monografia, Curso Técnico em Edificações, ETEC José Martimiano da Silva.

Sienge. (2025). Custo por etapa de obra: Como calcular e controlar seu orçamento. Accessed in: 23 ago. 2025: https://sienge.com.br/blog/custo-por-etapa-de-obra/.

Simon, H. A. The sciences of the artificial. MIT Press, 1996.

Sun, Y., Shi G., & Wu Q. (2011). Graphene based new energy materials. Journal Energy & Environmetal Science, 4, 1113–1114.

Teixeira, J. "Sustentabilidade: O que é, como funciona, benefícios e exemplos." 20 set. 2023. https://fia.com.br/blog/sustentabilidade/.

Tolmasquim, M. T. Fontes Alternativas de Energia no Brasil. Ed. Ciência Moderna. Rio de Janeiro, 2004. https://www.sienge.com.br/blog/custo-da-obra-por-etapa/ASM International. 2003.

Tripathi, S. N., Rao, G. S. S., Mathur, A. B., & Jasra, R. (2017). Polyolefin/Graphene nanocomposites: A review. RSC Advances, [S. l.], 7(38), 23615–23632.

World Commission on Environment and Development. (1987). Nosso Futuro Comum. 2° Edição. 23 ago. 2025: https://semil.sp.gov.br/educacaoambiental/prateleira-ambiental/nosso-futuro-comum-relatorio/.

Zeschky, Z., Winterhalter, S., & Gassmann, O. (2014) From cost to frugal and reverse innovation: Mapping the field and implications for global competitiveness. Research Technology Management, 57(4), 20–27, https://doi.org/10.5437/08956308X5704235.

9 Feasibility Study and Design of Heat Sink of Passive Cooling System

Jaswinder Singh Mehta

9.1 INTRODUCTION

Higher computational speed and lower processing time are the salient characteristics of today's electronic devices by virtue of recent technological advancements in this field. However, due to the miniaturization of these devices, they are also being subjected to higher thermal loads. Although many non-conventional techniques have evolved over years to provide the solution to the problem of increased heat flux, but reliability and suitability of these methods are of prime concern, as many times, external secondary systems such as pumps are required for the circulation of cooling medium that degrades the reliability of the main system.

Smart and passive cooling systems are thus the need of the hour so as to maintain the temperature of the system well below the safe working limit and to increase their service life. A ferrofluid-based cooling device can provide a potential solution to dissipate high heat flux (Yarahmadi et al., 2015; Aursand et al., 2016; Şeşen et al., 2012).

Ferrofluid consists of magnetic nanoparticles, usually in the range of up to about ten nanometers, uniformly dispersed in a carrier fluid. The typical composition of ferrofluid is shown in Figure 9.1.

The flow of ferrofluid can be manipulated just by changing the strength and direction of the magnetic field. In the absence of a magnetic field, nanoparticles in the ferrofluid remain randomly oriented in the base fluid, and the net magnetization of the fluid is zero. When ferrofluid is exposed to a non-uniform magnetic field, atomic moments of magnetic nanoparticles, which were originally randomly oriented, attempt to align along the direction of the applied magnetic field (Rosensweig, 1985). The degree of resulting magnetization in the fluid depends on the applied magnetic field strength and fluid temperature. When such a fluid is subjected to a temperature differential under the influence of an external magnetic field, a thrust force, called *Kelvin body force*, will be experienced by the fluid due to variation in the magnetization of the fluid, which leads to the setup of thermomagnetic convection in the fluid. Fluid column at lower temperature, thus, will exert a thrust on the hotter, lower magnetizing fluid that leads to flow of the fluid in the direction of the higher temperature section. This effect is similar to natural convection, where colder, more dense fluid displaces less dense fluid and convection currents are set up in the fluid (Singh Mehta

 DOI: 10.1201/9781003502647-11

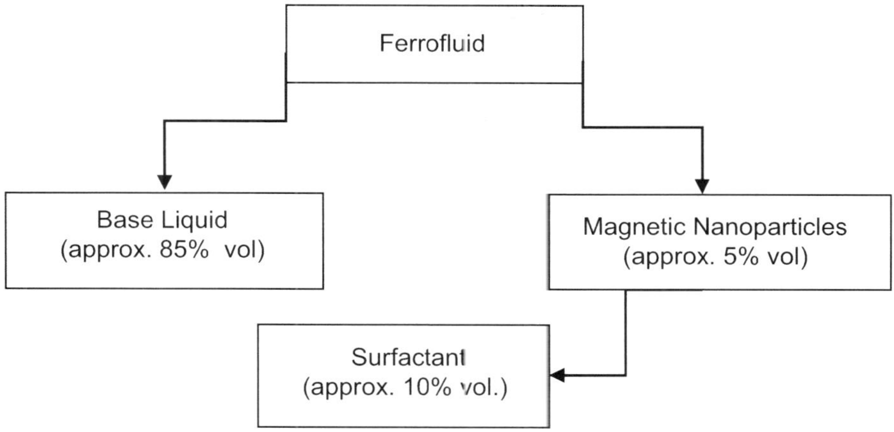

FIGURE 9.1 Composition of ferrofluid.

et al., 2018). Since the flow of such fluids is governed by the thermomagnetic convection principle, no external energy is required for its circulation. Ferrofluid-based cooling devices are thus totally passive, minimizing the energy requirement to run and thereby improving the reliability of the system.

An experimental evaluation of the performance of the cooling system for two different arrangements of permanent magnets was performed (Singh Mehta et al., 2018). COMSOL MultiPhysics was used to numerically analyze and generate the magnetic field distribution and the resulting magnetic force acting on the water-based ferrofluid flowing through a tube. Temperature measurements were recorded using the infrared thermography technique. Higher heat transfer augmentation and amplification of the Nusselt number were observed for a double in-line parallel arrangement of permanent magnets compared to a single-line arrangement.

An experimental evaluation of the flow behavior of hydrocarbon-based ferrofluid under a uniform magnetic field was done (Schwab et al., 1983). It was observed that, when the fluid layer was subjected to heat load, the onset of thermomagnetic convection resulted under the influence of a uniform vertical magnetic field at a critical Rayleigh number.

The lattice Boltzmann method was applied to numerically examine convection heat transfer of water-based ferrofluid in a multi-pipe heat exchanger (Ma et al., 2020). With an increase in Rayleigh number and nanoparticle volume fraction, energy transfer was found to enhance, while a reverse trend was observed with Hartmann number.

An experimental study was conducted to analyze the influence of intensity and direction of the magnetic field on the thermal performance of water-based ferrofluid (Sha et al., 2017). The heat transfer coefficient was observed to increase with a rise in temperature and volume concentration.

The effect of the shape of the channel and the magnetic field on thermomagnetic convection features of magnetite ferrofluid was observed (Lee & Kim, 2019). Different enclosure channel shapes, like square, separated square, circle, and separated circle, were subjected to a magnetic field generated by a permanent magnet. The interaction of nanoparticles with the magnetic field gradient, as represented by the magnetophoretic force, was found to increase with magnetic strength. Maximum convective heat flux was obtained for the circular enclosure channel, and the average temperature was also found to rise with an increase in heat flux.

A detailed comparison of the performance of the magnetic cooling system for water and kerosene-based ferrofluid was presented using COMSOL MultiPhysics (Garg et al., 2015). Measurements of temperature difference near the heat load region, pressure difference across the heating substrate, and heat transfer coefficient were taken for both water and kerosene-based ferrofluid. Smooth and sustainable flow was established for kerosene-based fluid, while pulsating flow with continuous flow variation was found for water-based ferrofluid.

Numerical investigation of the effect of gravitational induced sedimentation on heat transport and thermomagnetic convection in a kerosene-based ferrofluid was conducted (Jafari et al., 2015). Dimensionless parameters such as gravitational (Ra_g) and magnetic Rayleigh numbers (Ra_m) were employed to evaluate the thermal efficiency of the system. It was concluded from the study that if the ratio of magnetic energy to gravitational energy is maintained above unity, the probability of formation of agglomerates is negligible. Also, heat transfer by thermomagnetic convection was found to be higher in comparison to pure natural convection.

The effect of magnetic field strength and nanoparticle concentration on velocity, temperature, flow drag, pressure drop, and heat transfer rate of water-based magnetite nanofluid was analyzed under the influence of electromagnetic field (Lo & Weng, 2015). The finite difference method was used to numerically investigate the thermal flow fields. It was revealed from the study that more be nanoparticle concentration, the higher the probability of clogging, leading to greater pressure drop and a decrease in average heat transfer rate. The strength of the magnetic field also had a negative impact, resulting in the generation of a higher pressure drop and a higher average flow drag that led to an insignificant rise in the average heat transfer rate.

The present study deals with performing a feasibility check of the working of a passive cooling device employing ferrofluid as a cooling medium and determining the number and height of annular fins for cooling of the ferrofluid in the closed channel using COMSOL MultiPhysics. The ferrofluid was found to circulate on its own in the closed loop without any assistance of external energy in the presence of a magnetic field generated by the permanent magnet, and the temperature drop across the heat sink was observed to increase with the number of fins. Fin height also contributes to the temperature drop, but, beyond a certain value, its effect was found to be insignificant.

9.2 MATHEMATICAL MODELING AND NUMERICAL SIMULATION

9.2.1 Geometrical Model

A schematic of a two-dimensional geometric model used for conducting the feasibility study of the flow of ferrofluid in the oval-shaped closed loop is shown in Figure 9.2. A permanent magnet was positioned near the heat source to provide the

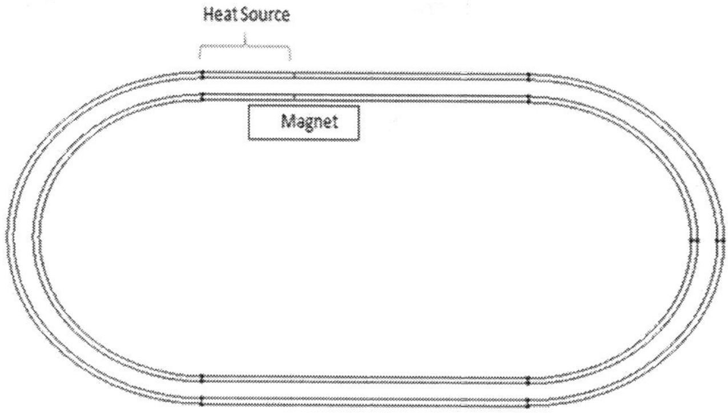

FIGURE 9.2 Two-dimensional geometrical loop.

necessary thrust force required for the movement of the fluid as per the thermomagnetic convection principle. A line heat source was employed to provide the heat input.

9.2.2 Equations Governing Flow of Ferrofluid

The following governing equations (Rosensweig, 1985) were used to simulate the flow of ferrofluid through the closed loop:

Continuity Equation

$$\frac{\partial \rho}{\partial t} + \nabla.\left(\rho\vec{u}\right) = 0 \quad \text{(IX-1)}$$

Momentum Equation

$$\rho\partial\vec{u} / \partial t + \rho\vec{u}.\nabla\vec{u} = -\nabla p - \nabla.(\mu(\nabla\vec{u} + \left(\nabla\vec{u}\right)^{T}) + \vec{F} \quad \text{(IX-2)}$$

Energy Equation

$$\rho c_p\left(\frac{\partial T}{\partial t} + \vec{u}.\,\nabla T\right) = k\nabla^2 T \quad \text{(IX-3)}$$

Magnetic Induction

$$\vec{B} = \mu_o\left(\vec{H} + \vec{M}\right) \quad \text{(IX-4)}$$

Thrust Force (Kelvin Body Force)

$$\vec{F} = \left(\vec{M}.\nabla\right)\vec{B} \quad \text{(IX-5)}$$

9.2.3 Boundary and Initial Conditions

i. The fluid was considered as laminar and incompressible in nature.
ii. At the inner wall surface, zero slip conditions were assumed for the fluid flow.
iii. Surrounding fluid and initial fluid temperature at the start of the computational study were taken to be the same as 273.15 K.
iv. The viscous dissipation effect was neglected.

Table 9.1 presents the specifications and other properties of the ferrofluid used in the present investigation.

9.2.4 Meshed Model and Statistics

The meshed model was created using the physics-controlled mesh option available in COMSOL MultiPhysics, as demonstrated in Figure 9.3 with statistics presented in Table 9.2 for the current study.

9.3 RESULTS AND DISCUSSION

The feasibility study was performed first by observing the flow behavior of the fluid in the absence and thereafter in the presence of a magnetic field. Fluid movement was

TABLE 9.1
Kerosene-Based Ferrofluid Properties

S. No.	Fluid Parameters	Value
1	Ferrofluid viscosity, μ	1.5 cP
2	Density, ρ	900 kg/m^3
3	Conductivity, k	0.178 W/(m-K)
4	Curie point of fluid	45°C
5	Magnetic susceptibility	0.386

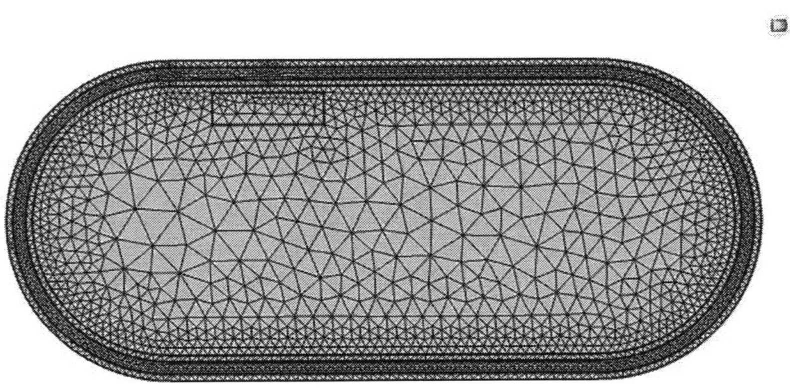

FIGURE 9.3 Meshed model of the geometry.

TABLE 9.2
Mesh Statistics

S. No.	Property	Value
1	Minimum element quality	0.1214
2	Number of elements	24,525
3	Number of vertex elements	376
4	Number of boundary elements	3561

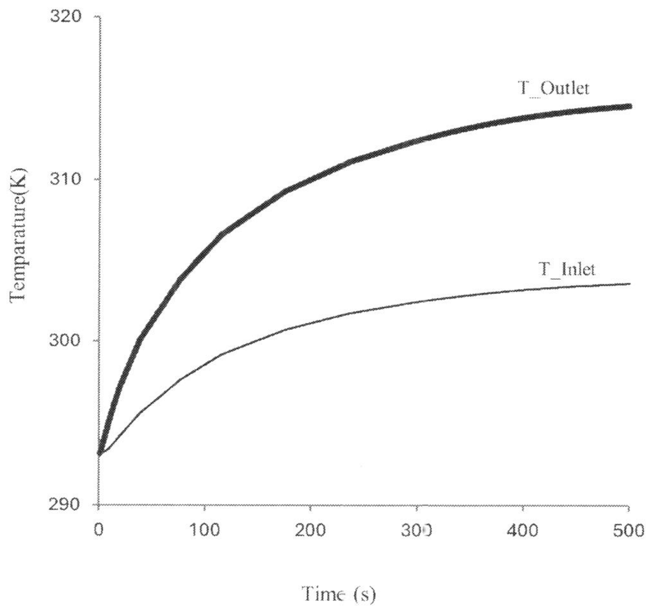

FIGURE 9.4 Temperature–time curves for the fluid (under a magnetic field).

not noticed in the absence of a magnetic field. However, when the fluid was subjected to an external magnetic field generated by a permanent magnet, the temperature curves at two points, that is, upstream and downstream of the heater, tend to separate from each other and flatten out after a short while. The temperature differential led to non-uniform magnetization in the fluid, as a result of which the fluid experiences a thrust force and flow of fluid initiated in a clockwise direction owing to thermo-magnetic convection, as presented in Figure 9.4. The movement of the fluid without the aid of any external device thus confirms the working of the cooling system on its own.

As the fluid passes through the heat source portion, it extracts heat and thus requires to be cooled down so that the coolant is available for the next cycle of operation. The heat sink portion is thus an important element of the system. The temperature of the fluid should drop as it flows through the heat sink region. An effective way to promote a higher heat transfer rate in the heat sink region is by increasing the surface area, and fins are the obvious choice whenever the objective is to increase the heat dissipation from the surface. Annular fins of uniform thickness are attached to the opposite leg of the loop, as shown in Figure 9.5. Numerical analysis was done by varying the parameters such as the number of fins and the height of the fins and evaluating the effect of these parameters on temperature drop across the fins using COMSOL MultiPhysics, as presented in Figures 9.6 and 9.7.

Numerical simulation, as performed earlier, was conducted again by taking the height of the fin as 6 mm and varying the number of fins.

The initial conditions used for the simulation work were kept the same during this stage also. COMSOL MultiPhysics was used to simulate the fluid flow and temperature contours. Close-up of the fin area has been captured for each model, and temperature contours have been shown in Figure 9.7.

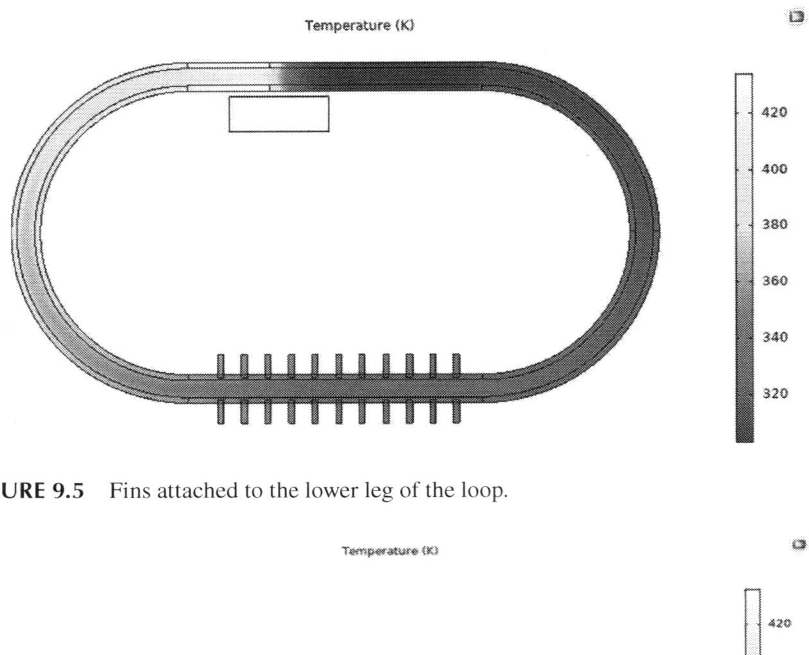

FIGURE 9.5 Fins attached to the lower leg of the loop.

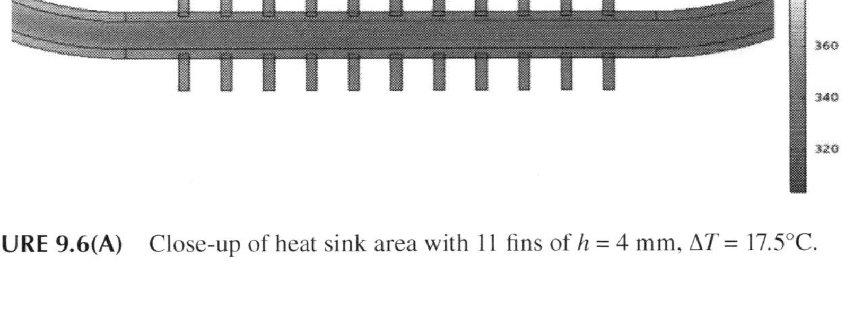

FIGURE 9.6(A) Close-up of heat sink area with 11 fins of $h = 4$ mm, $\Delta T = 17.5°C$.

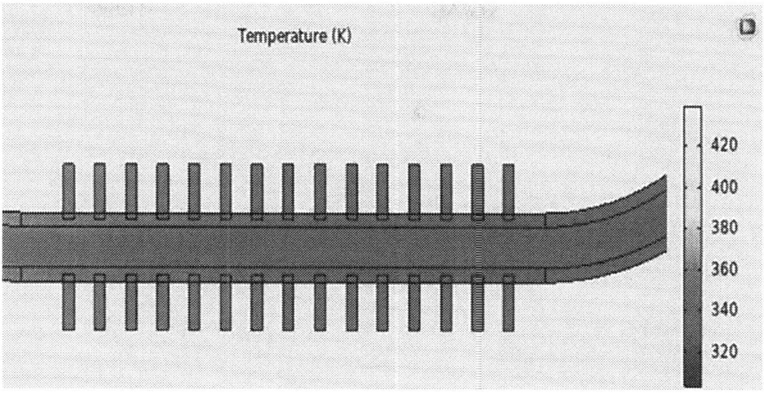

FIGURE 9.6(B) Close-up of heat sink area with 15 fins of $h = 4$ mm, $\Delta T = 22°C$.

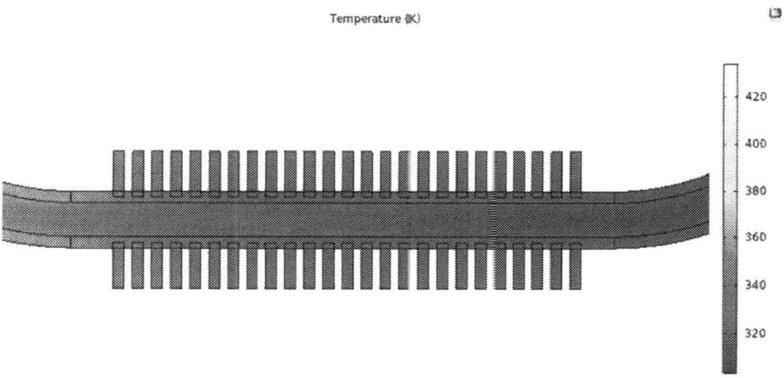

FIGURE 9.6(C) Close-up of heat sink area with 21 fins of $h = 4$ mm, $\Delta T = 35°C$.

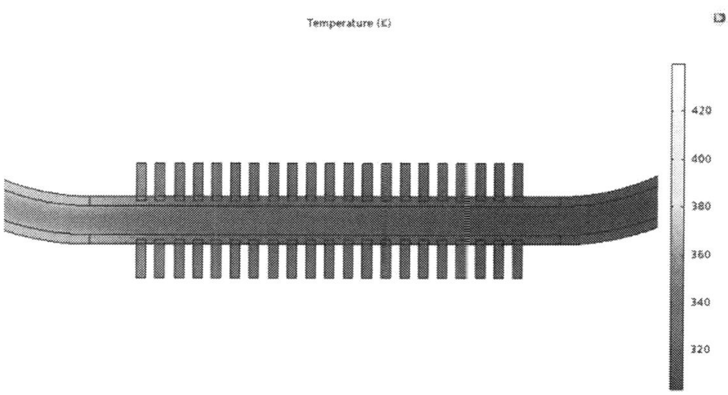

FIGURE 9.6(D) Close-up of heat sink area with 25 fins of $h = 4$ mm, $\Delta T = 39°C$

FIGURE 9.6(A–D) Temperature distribution in heat sink region for varying number of fins with fin height = 4 mm.

Figure 9.7(a–d) shows the temperature distribution in the heat sink region for varying numbers of fins with a fin height set at 6 mm.

Figure 9.8 summarizes the simulation results as obtained by varying the number and height of the fins on temperature drop across the fins (ΔT).

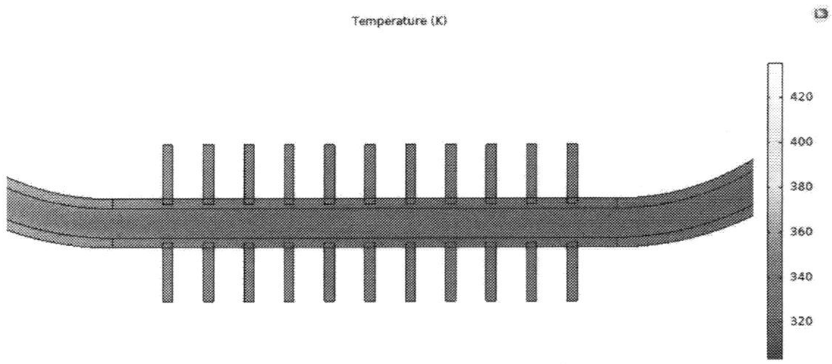

FIGURE 9.7(A) Close-up of heat sink area with 11 fins of $h = 6$ mm, $\Delta T = 22$°C.

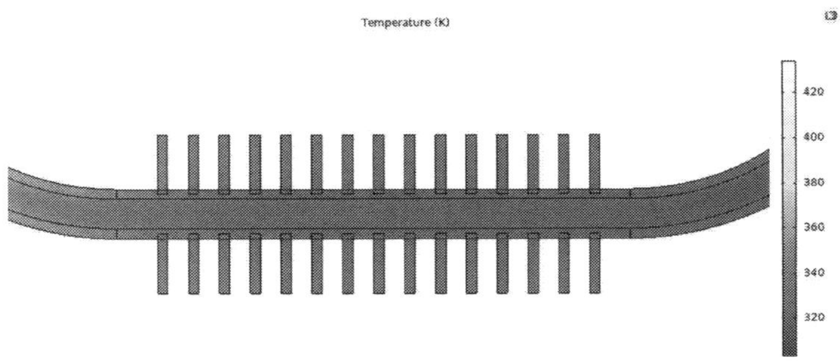

FIGURE 9.7(B) Close-up of heat sink area with 15 fins of $h = 6$ mm, $\Delta T = 26$°C.

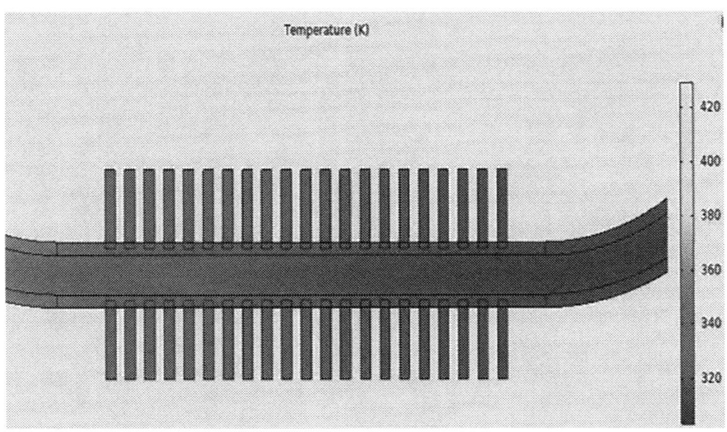

FIGURE 9.7(C) Close-up of heat sink area with 21 fins of $h = 6$ mm, $\Delta T = 40$°C.

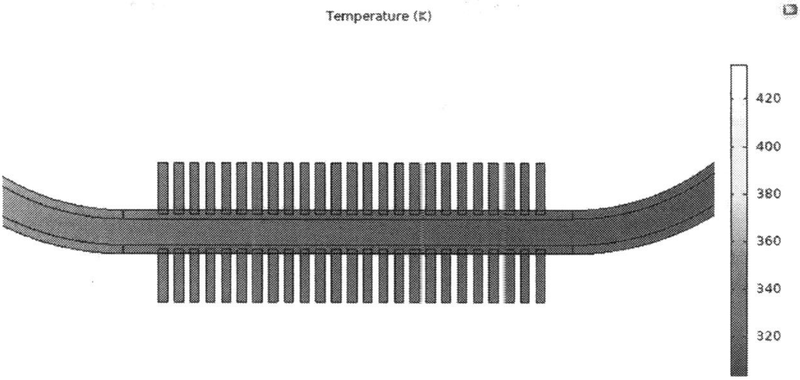

FIGURE 9.7(D) Close-up of heat sink area with 25 fins of $h = 6$ mm, $\Delta T = 42°C$.

FIGURE 9.7(A–D) Temperature distribution in heat sink region for varying number of fins with fin height, $h = 6$ mm.

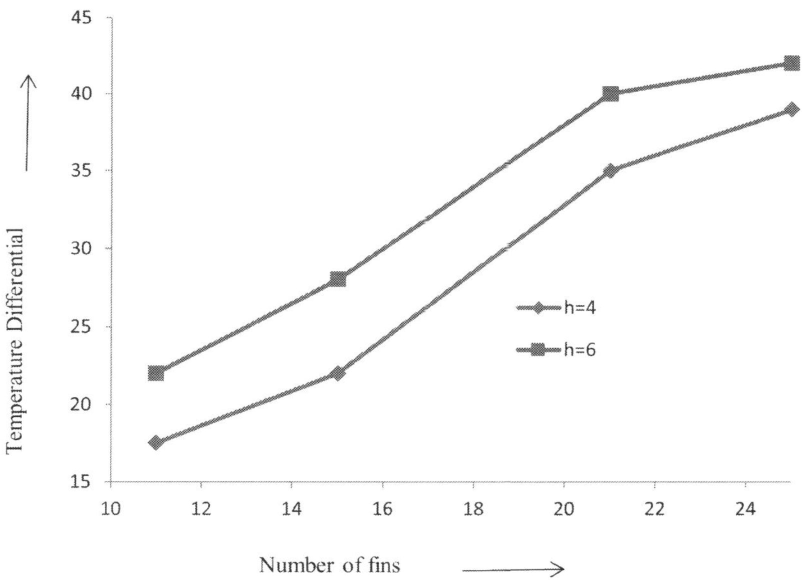

FIGURE 9.8 Variation of temperature differential across the fins with the number of fins and height of fins.

9.4 CONCLUSION

The following inferences can be drawn from the study conducted in this work:

i. The magnetic field acts as the driving force for the fluid. In the absence of a magnetic field, ferrofluid movement was not observed. However, upon application of a magnetic field, a temperature differential was set up in the fluid near the heat source region, as a result of which the fluid starts to flow in the loop. Flow was established on its own without the necessity of any external energy source, and the cooling device worked satisfactorily.

ii. The study further finds the effect of the number of fins and the height of fins on the temperature drop across the heat sink region. It was revealed that as the number of fins increases from 10 to 25, the temperature differential across the fins also increases. However, the temperature drop becomes smaller with an increasing number of fins. Also, the height of the fins has a direct bearing on the temperature drop across the fins. But the two curves tend to converge as the number of fins increases, revealing that the height of the fins will have an insignificant effect on the heat transfer beyond a certain value.

REFERENCES

Aursand, E., Gjennestad, M. A., Lervåg, K. Y., & Lund, H. (2016). A multi-phase ferrofluid flow model with equation of state for thermomagnetic pumping and heat transfer. *Journal of Magnetism and Magnetic Materials*, *402*, 8–19. https://doi.org/10.1016/j.jmmm.2015.11.042.

Garg, H., Negi, V. S., Kharola, A. S., & Garg, N. (2015). Numerical analysis of heat transfer parameters using ferrofluid. *Nanoscale and Microscale Thermophysical Engineering*, *19*(4), 258–271. https://doi.org/10.1080/15567265.2015.1093893.

Jafari, A., Shahmohammadi, A., & Mousavi, S. M. (2015). CFD investigation of gravitational sedimentation effect on heat transfer of a nano-ferrofluid. *Iranian Journal of Chemistry and Chemical Engineering*, *34*(1), 87–96. https://doi.org/10.30492/IJCCE.2015.12686.

Lee, M., & Kim, Youn-Jea. (2019). Thermomagnetic convection of ferrofluid in an enclosure channel with an internal magnetic field. *Micromachines*, *10*, 553. https://doi.org/10.3390/mi10090553.

Lo, K. J., & Weng, H. C. (2015). Convective heat transfer of magnetic nanofluids in a microtube. *Smart Science*, *3*(2), 56–64. https://doi.org/10.1080/23080477.2015.11665637.

Ma, Y., Mohebbi, R., Rashidi, M. M., Yang, Z., & Sheremet, M. (2020). Nanoliquid thermal convection in I-shaped multiple-pipe heat exchanger under magnetic field influence. *Physica A: Statistical Mechanics and its Applications*, *550*, 124028. https://doi.org/10.1016/j.physa.2019.124028.

Rosensweig, R. E. (1985). Directions in ferrohydrodynamics. *Journal of Applied Physics*, *57*(8), 4259–4264. https://doi.org/10.1063/1.334579.

Schwab, L., Hildebrandt, U., & Stierstadt, K. (1983). Magnetic bénard convection. *Journal of Magnetism and Magnetic Materials*, *39*(1–2), 113–114. https://doi.org/10.1016/0304-8853(83)90412-2.

Şeşen, M., Tekşen, Y., Şendur, K., Pınar Mengüç, M., Öztürk, H., Yağcı Acar, H. F., & Koşar, A. (2012). Heat transfer enhancement with actuation of magnetic nanoparticles suspended in a base fluid. *Journal of Applied Physics*, *112*(6). https://doi.org/10.1063/1.4752729.

Sha, L., Ju, Y., & Zhang, H. (2017). The influence of the magnetic field on the convective heat transfer characteristics of Fe3O4/water nanofluids. *Applied Thermal Engineering*, *126*, 108–116. https://doi.org/10.1016/j.applthermaleng.2017.07.150.

Singh Mehta, J., Kumar, R., Kumar, H., & Garg, H. (2018). Convective heat transfer enhancement using ferrofluid: A review. *Journal of Thermal Science and Engineering Applications*, *10*(2), 020801. https://doi.org/10.1115/1.4037200.

Yarahmadi, M., Goudarzi, H. M., & Shafii, M. B. (2015). Experimental investigation into laminar forced convective heat transfer of ferrofluids under constant and oscillating magnetic field with different magnetic field arrangements and oscillation modes. *Experimental Thermal and Fluid Science*, *68*, 601–611. https://doi.org/10.1016/j.expthermflusci.2015.07.002.

10 Effects of the Addition of Plastic Waste on the Strength Properties of Concrete and Bituminous Asphalt

Ranjodh Singh and Ravpreet Kaur

10.1 MATERIAL AND PROPERTIES

10.1.1 Polythene Bags Used in Daily Life

The 1960s saw the introduction of the polyethylene or polypropylene plastic bag, which is seen as a symbol of the consumer society. Around the world, more than 600 billion plastic bags are produced annually; the average European uses about 500 of them. When compared to how long they last, they are often only used once. Plastic bags, in particular, thin ones, present a serious ecological problem. But the widespread use of plastic also brings up a number of problems. Huge amounts of plastic can be disposed of with ease. Because typical plastic is not biodegradable, improperly disposed of plastic garbage harms our ecosystem for decades or even centuries.

10.1.2 Multilayer Plastic Used in Food Packaging

The packaging used in the food industry greatly affects the shelf life of food. It helps to preserve food safety and provides protection when handling, moving, and presenting food. The movement of gases and vapors between the environment and the food can be encouraged or inhibited by the container. The name for this gas or vapour is permeant. Limiting the amount of some permeants in a product lengthens its shelf life by minimizing biological and chemical deterioration, microbial growth, and modifications to the food's color, flavor, texture, and odor. In recent decades, there has been a marked growth in the use of plastic packaging. Due to their flexibility, low density, processability, and adaptability, plastics frequently turn out to be more environmentally friendly than conventional packaging materials after a complete life cycle assessment. The global use of plastics has increased enormously over the past few decades, catching up to the consumption of steel and paper. The longest period

 DOI: 10.1201/9781003502647-12

of time between a food's production and its intended use or consumption is referred to as its shelf life. Foods that have undergone any chemical, biological, or physical modification will have this time reduced. These changes could alter how the consumer perceives foods in terms of their sensory aspects, nutritional value, or safety, which could put the consumer at risk.

10.1.3 Concrete

It is a material that is used in construction and is made up of cement, sand, and both fine and coarse aggregates. This combination is frequently used in construction since it gets harder over time. The most popular type of cement is Portland cement, which is used to make concrete. During the construction of buildings and roads, this concrete is used to make a wide range of products, including slabs, beams, columns, foundations, and other parts of rigid, load-bearing pavements. Commonly used concrete grades include M15, M20, and M25. Concrete of grade M15 plain cement is used as a base or for leveling.

10.1.4 Bitumen Asphalt

It is a graded mixture of aggregate, filler, and bitumen. It is used to construct operational surfaces, like as runways, taxiways, aprons, and road pavements, and acts as the crust's top layer to enable efficient traffic movement.

According to several laboratory or cylinder tests, mixed design concrete is the one that completes the mixing ratio because of its compressive strength. This process is frequently referred to as blending design. These tests are conducted to identify the best mix of locally available materials to establish the necessary strength in line with the structural design. A blended design offers a productive use of ingredients. Various physical properties of cement used are mentioned in Table 10.1. Tables 10.2 and 10.3 show various physical properties of fine and coarse aggregates, respectively. The origin and source of different waste plastics used are presented in Table 10.4.

10.2 ADVISORY RELATED TO THE USE OF DIFFERENT PLASTIC WASTES IN CONSTRUCTION AND MAINTENANCE OF PRADHAN MANTRI GRAM SADAK YOJANA ROADS UNDER SWACHHATA HI SEVA

The Honorable Prime Minister's Independence Day address from August 15, 2019, which covered how to handle plastic debris. is referred to in this. According to the government, this year's Swachhata Hi Seva would be centered on managing plastic trash. From September 11 to October 27, 2019, the government will observe the plastic waste free campaign. The program's goal is to manage plastic waste more quickly in towns, cities, and public places like schools, hospitals, mandies, and places of worship. To increase the awareness of the use of single-use plastic among both the rural and urban population, the Secretary of Rural Development of the Government of India has already written to the Chief Secretaries of the States via D.O. Letter No. L-12046/03/2019-SAGY dated September 3, 2019.

TABLE 10.1
Physical Properties of Cement

S. No.	Characteristics	Units	Results Obtained	Value Specified
1.	Fineness (specific surface)	cm^2/g	3000	3500
2.	Specific gravity	gm/cm^3	3.15	
3.	Normal consistency of cement (% of cement by weight)	%	29	30
4.	Setting time			
	(i) Initial	minutes	135	30 (minimum)
	(ii) Final	minutes	225	600 (maximum)

TABLE 10.2
Physical Properties of the Fine Aggregates

S. No.	Characterstics	Results
1	Specific gravity	2.62
2	Fineness modulus	2.86
3	Water absorption %	1.8
4	Bulk density loose (kg/m^3)	1680
5	Free surface water	.
6	Zone	II

TABLE 10.3
Physical Properties of the Coarse Aggregates

S. No.	Characteristics	Results Obtained	
		20 mm	10 mm
1	Specific gravity	2.67	2.62
2	Water absorption %	0.48	0.56
3	Free surface water	.	.
4	Aggregate impact value	10.49%	11.61%
5	Flakiness index	6.8%	10.00%
6	Elongation index	17.6%	20.8%

TABLE 10.4
Waste Plastic and Its Source

Waste Plastic	Origin
Low-density polyethylene (LDPE)	Carry bags, sacks, milk pouches, bin lining, detergent bottles
High-density polyethylene (HDPE)	Carry bags, bottle caps, and household articles
Polyethylene terephthalate (PET)	Drinking water bottles
Polypropylene (PP)	Bottle caps and closures, wrappers of detergents, biscuits and wafer packets, microwave trays for readymade meals
Polystyrene (PS)	Yogurt Pots, clear egg packs, bottle caps, food trays
Polyvinyl chloride (PVC)	Mineral water bottles, credit cards, toys, electrical fittings, furniture, folders, and pens

FIGURE 10.1 Waste plastic single layer (Raghatate Atul & Saikia, 2013).

FIGURE 10.2 Multilayer plastic (Raghatate Atul & Saikia, 2013).

1. The Ministry of Rural Improvement has been tasked with, among other things, utilizing the plastic waste acquired throughout the campaign for road improvement as part of the Pradhan Mantri Gram Sadak Yojana (PMGSY). The PMGSY program's efforts to construct rural roads using recycled plastic as an environmentally friendly way have received major contributions from numerous States. As part of the PMGSY, more than 13,000 km of roads have already been constructed nationally using waste plastic. Additionally, it has been demonstrated that these roads operate quite well.

2. According to research, waste plastics hold great promise for use in bituminous construction because, even in small quantities, such as 6% to 8% by weight of bitumen, they greatly enhance the substance. The bituminous mix's Marshall Stability, strength, fatigue life, stronger resistance to deformation, higher resistance to water-induced damages, and other required qualities make it possible for improved lifetime and pavement performance. As a result, recycling plastic debris during road construction helps create ecologically friendly roads while simultaneously boosting the economy. Utilizing waste plastic alone in OGPC can reduce construction costs by between Rs. 10,000 and 25,000 per kilometer, according to bitumen and waste plastic rates. A training film on the use of plastic wastes in the construction of bituminous roads was prepared by NRIDA and is available for download from the homepage of pmgsy.nic.in.
3. I respectfully request your personal support in promoting the usage of plastic waste in the construction of and its maintenance. The Ministry/NRIDA would offer all support with training requirements, capacity development, or handholding for this goal.

10.2.1 Plastic Waste Roads in Punjab

Although there is no code for the use of multilayer plastic, the overall length of Premix Carpet installed in Punjab using single-layer plastic waste is 1131.85 km and the entire length using multilayer plastic is 325.64 km, according to IRC SP:98:2013 requirements.

10.3 REVIEW OF LITERATURE

To determine the effects of adding polythene fiber (domestic waste) at a dose of 0.5% by weight of cement, Kandasamy and Murugesan (2011) conducted an investigation. Two of the qualities being studied are compressive strength and flexural strength. The tests were done on an M20 mix and carried out in accordance with the recommended procedures of the relevant regulations. Comparisons between the results and conclusions are made.

According to Raghatate Atul and Saikia (2013), using plastic bags to transport vegetables, meat, and other items causes a severe environmental issue. Plastic bags can last in the environment for up to 1000 years, and, because of this, the total number of plastic bags used each year rises. Large-scale plastic bag disposal could contaminate the air, water, and land. The suggested concrete, which is created by incorporating plastic into concrete, may assist in reusing plastic bags as a component material of concrete, improving particular features of concrete.

As part of Luis Ferreira et al.'s (2012) study on the effects of the curing environment on the mechanical parameters of concrete, several mixes were created with replacement ratios of 0%, 7.5%, and 15% of natural aggregates by three types of plastic aggregate. The mechanical performance of concrete was evaluated in three different curing environments—lab settings, a wet chamber, and the outdoors—to represent the many scenarios that concrete may run into. Concrete's elastic modulus and compressive and breaking tensile strengths drop when plastic is added and as its size grows. But it improved its abrasion wear resistance. In damper conditions, plastic-free concrete fared best. Additionally, it was shown that different curing regimes had different effects on how concrete reacts to the addition of plastic.

Costa et al. (2012) noted a sharp rise in plastic use over the past few years, which has produced massive volumes of waste related to plastic. Given its benefits to the environment and the economy, recycling plastic trash to create new materials like concrete or mortar seems to be one of the greatest ways to get rid of plastic garbage. Many studies have been conducted or are currently being conducted to assess the properties of cement composites that use different forms of plastic waste as aggregate, filler, or fiber. This study investigates the use of recycled plastic trash as aggregate in the creation of cement mortar and concrete.

To create self-compacting mortars (SCMs), polyethylene terephthalate (PET), which is used to make plastic bags, can be recycled and used as a fine aggregate in place of sand, according to Brahim Safi et al. (2013). The physical and mechanical characteristics of the SCMs made of plastic waste were investigated experimentally. Sand is substituted with plastic waste at weight ratios of 0%, 10%, 20%, 30%, and 50%. A secondary investigation on the micro-structure of the interface of cementitious matrix and plastic waste was carried out in addition to evaluating the physical (bulk density, porosity, water absorption, and ultrasonic pulse velocity testing) and mechanical (bulk compressive and flexural strength) properties of SCMs.

The goal of this study by Raghatate Atul and Saikia (2013) was to assess the impact of recycled PET aggregate's size and shape on concrete's initial and hardened qualities, including its ability to resist abrasion. Three different-shaped and sized PET aggregates were substituted in the concrete mixes for the natural aggregate proportions of 5%, 10%, and 15%. The results demonstrate that adding PET aggregate in the form of pellets slightly improves the slump of fresh concrete. Fresh concrete's slump is significantly reduced by a flakier plastic aggregate, and the slump is further reduced as the volume and size of this sort of PET aggregate increase. Due to the inclusion of PET aggregate, the compressive strength, tensile splitting strength, elastic modulus, and flexural strength of concrete diminish, and the deterioration of these properties worsens with increasing quantity of this aggregate.

Enhancing the resulting binders' qualities for use in high-performance asphalt mixes, Costa et al. (2013) analyze the possible advantages of changing the bitumen with different plastic wastes, such as crumb rubber, acrylonitrile–butadiene–styrene, ethylene–vinyl acetate, and polyethylene (high-density polyethylene and low-density polyethylene (LDPE)). The structural and functional performance of road pavements must be enhanced since increased traffic places more strain on them. Polymers can be added to bitumen or asphalt mixtures as a way to significantly improve the quality of flexible pavements. While bitumen can be altered with new polymers to improve the characteristics of asphalt mixtures, recycled plastic can also achieve the same results while also having extra environmental advantages.

The compressive strength of concrete built with plastic bags as a fibrous material was compared in a study by Kumar et al. (2014). The study focuses on the impact of polyethylene plastic bags on the workability and compressive strength of M25 concrete, and solid waste management systems around the globe face the difficult challenge of safely disposing of non-recyclable thin polythene bags. In many nations, concrete is the material of choice for construction. We can use a lot of plastic garbage by adding it to concrete mass if plastic wastes can be put into concrete mass in dry form without significantly changing its fundamental features or weakening it.

Huge amounts of plastic in diverse forms and types were examined by Merbouh and Glaoui (2015). There are significant issues with these wastes in nature, in urban areas, and in the countryside that raise concerns about growing problems with hygiene and the environment. Concrete made of cement is a compact, waterproof substance that can prevent environmental pollution by withstanding the negative effects of plastic. It is also reasonably innocuous and durable. As a method of cost-effective recovery and reuse, it was deemed appropriate in this context to incorporate plastic waste (bag: LDPE) and plastic trash (bottle: PVC) into the concrete composition. These materials may be used to create parts that need less performance, such as tiny works, architectural concrete, interior ornamental elements, infill, surface coverage, and roofs.

10.4 TESTING PROCEDURE

The following flow chart shows the methodology of the project plastic waste used in concrete.

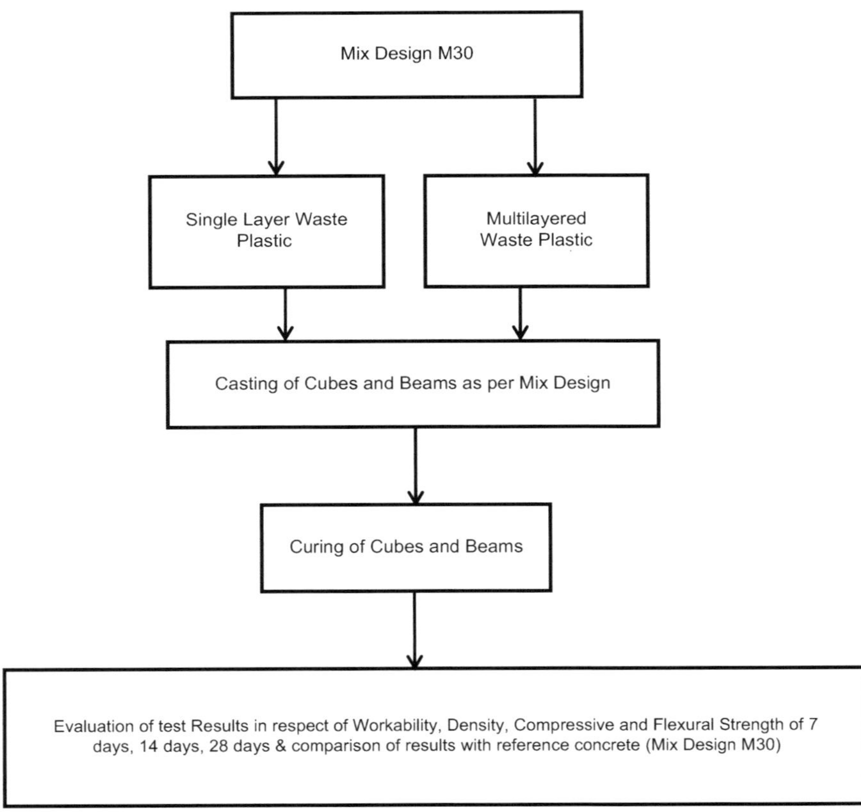

The following flow chart shows the methodology of the project plastic waste used in bitumen asphalt.

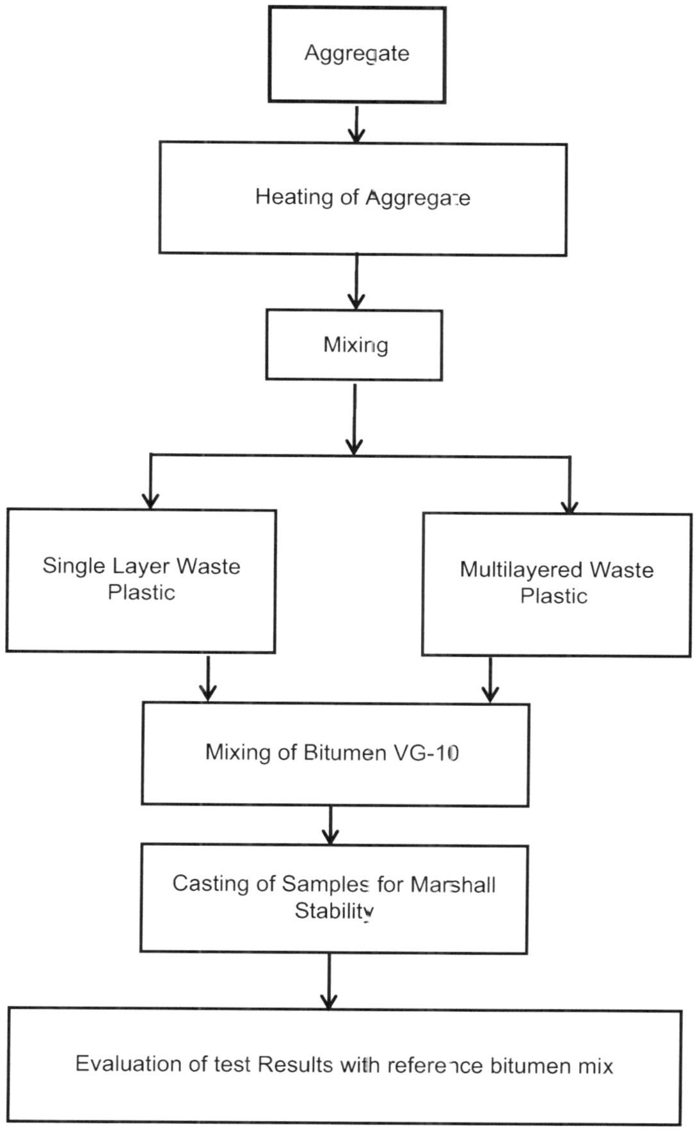

After samples for WP1, WP2, WP3, WP4, WP5, WP6, and WP7 and MLP1, MLP2, MLP3, MLP4, MLP5, MLP6, and MLP7 were cast and tested at the curing period of 7, 14, and 28 days for all of the strength tests as described earlier, M1 conventional mix was then used. After getting the results of all the samples, comparison

TABLE 10.5
Quantities of Material per Cubic Meter of Concrete (Merbouh & Glaoui, 2015)

						Coarse Aggregate (kg)		
Mix Designation	% of Waste Plastic	Wt. of Waste Plastic (kg)	Cement (kg)	Water (kg)	Admixture (kg)	20 mm	10 mm	Fine Aggregate (kg)
M_1	–	–	390	148	5.46	565	691	675
WP_1	0.10	3.90	390	148	5.46	565	691	675
WP_2	0.20	7.80	390	148	5.46	565	691	675
WP_3	0.30	11.70	390	148	5.46	565	691	675
WP_4	0.40	15.60	390	148	5.46	565	691	675
WP_5	0.5	19.5	390	148	5.46	565	691	675
WP_6	0.75	29.25	390	148	5.46	565	691	675
WP_7	1.00	39	390	148	5.46	565	691	675
MLP_1	0.110	3.900	390.0	148.0	5.460	565	691	675
MLP_2	0.200	7.800	390.0	148.0	5.460	565	691	675
MLP_3	0.300	11.700	390.0	148.0	5.460	565	691	675
MLP_4	0.400	15.600	390.0	148.0	5.460	565	691	675
MLP_5	0.500	19.500	390.0	148.0	5.460	565	691	675
MLP_6	0.750	29.250	390.0	148.0	5.460	565	691	675
MLP_7	1.00	39.0	390.0	148.0	5.460	565	691	675

of the values of both plastic waste materials and conventional concrete, we establish the relation between the obtained results. The notation used for different concrete mixes is shown here:

Description

Mix Designation	% of Waste Plastic
M_1	Concrete mix with 0% waste plastic
WP_1	Concrete + 0.1% waste plastic single layer
WP_2	Concrete + 0.2% waste plastic single layer
WP_3	Concrete + 0.3% waste plastic single layer
WP_4	Concrete + 0.4% waste plastic single layer
WP_5	Concrete + 0.5% waste plastic single layer
WP_5	Concrete + 0.75% waste plastic single layer
WP_7	Concrete + 1.00% waste plastic single layer
MLP_1	Concrete + 0.1% multilayer waste plastic
MLP_2	Concrete + 0.2% multilayer waste plastic
MLP_3	Concrete + 0.3% multilayer waste plastic
MLP_4	Concrete + 0.4% multilayer waste plastic
MLP_5	Concrete + 0.5% multilayer waste plastic
MLP_6	Concrete + 0.75% multilayer waste plastic
MLP_7	Concrete + 1.00% multilayer waste plastic

TABLE 10.6
Quantities of Material of Bitumen Mix for Molds (Raghatate & Saikia, 2013)

Sample	Percentage of Waste Plastic by Weight of Bitumen	Coarse Aggregate 11.2 mm Gauge (g)	Bitumen VG-10 (g)	Waste Plastic: Single Layer (g)	Waste Plastic: Multiplayer (g)
BM	0	1150	54.18	0	0
BM_1	7	1150	50.39	3.79	0
BM_2	7	1150	50.39	0	3.79

Comparison of Marshall Stability values and flow values has been done, using the same waste plastic, i.e. 7% by replacing the bitumen content in 4.5% of the bitumen mix with nominal mix as per guidelines of IRC:SP:98-2013.

Description

Mix Designation	% of Waste Plastic
BM	Bitumen VG-10 without plastic waste
BM_1	Bitumen VG-10 + @7% waste plastic replacing bitumen
BM_2	Bitumen VG-10 + @7% multilayer waste plastic replacing bitumen

(A)

(B)

FIGURE 10.3 (A) AND (B) Specimens (Kumar et al., 2014).

10.5 RESULTS AND CONCLUSIONS

Results for different tests conducted for different mixes are given as under

Discussion

1. The results show that for up to 0.1% waste plastic there is less change in the compressive strength. However, after 0.1% the compressive strength begins to decline, and with 1% waste plastic, the decrease in compressive strength is 19%. The compressive strength also begins to decline quickly after 0.2%

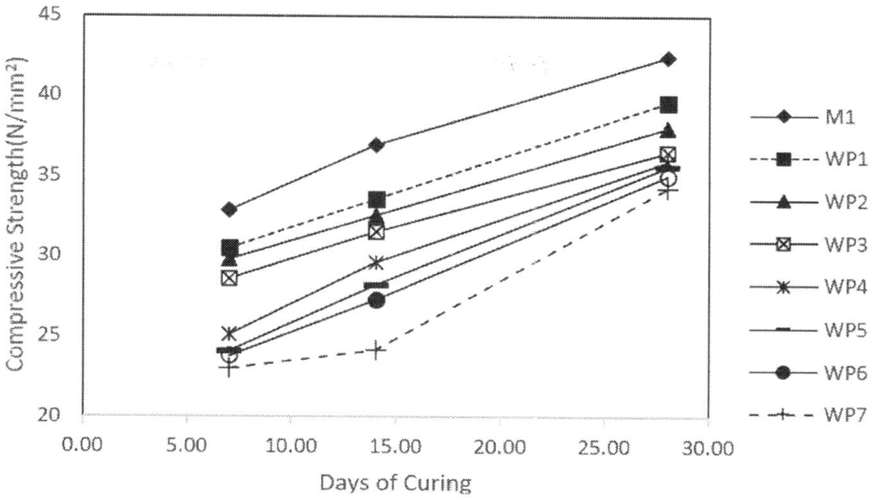

FIGURE 10.4 Comparison of compressive strength for single-layer plastic waste mix after 7, 14, and 28 days of curing (Kumar et al., 2014).

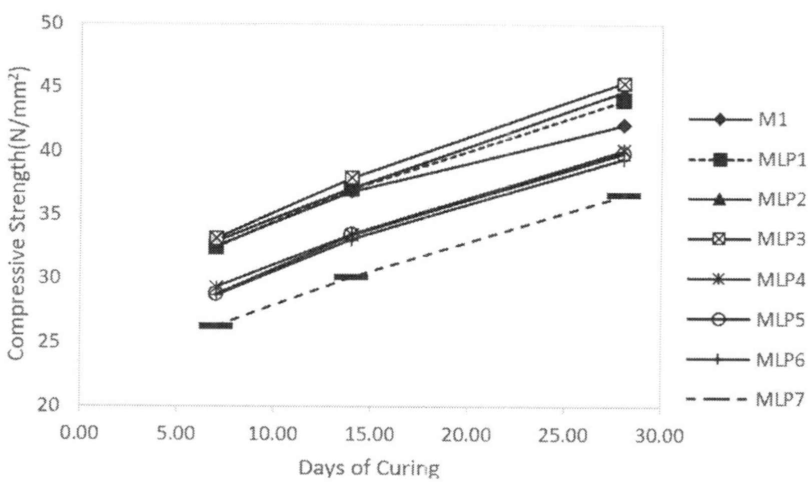

FIGURE 10.5 Comparison of compressive strength for multilayer plastic waste mix after 7, 14, and 28 days of curing (Raghatate & Saikia, 2013).

waste plastic single layer, whereas, with the addition of MLP waste, the compressive strength increases up to 0.3% and then declines, and with 1% MLP waste the decrease in compressive strength is 13.1%.

2. Factors responsible for low compressive strength are the hydrophobic nature of plastic waste, which can inhibit cement hydration reaction by restricting water movement.

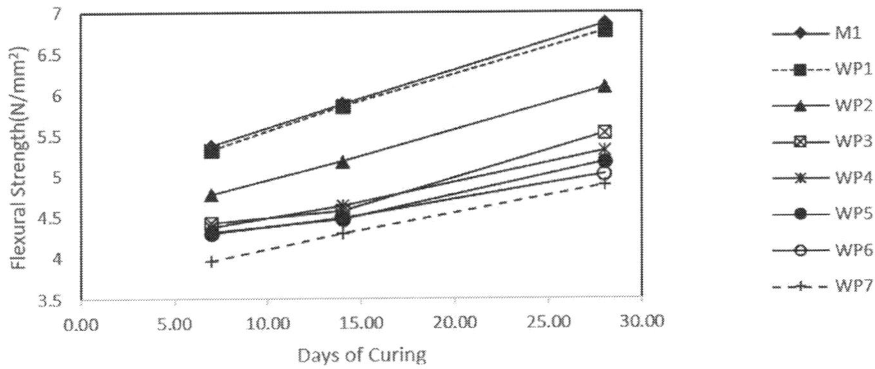

FIGURE 10.6 Comparison of flexural strength for single-layer plastic waste mix after 7, 14, and 28 days of curing (Bhogayata & Arora, 2017).

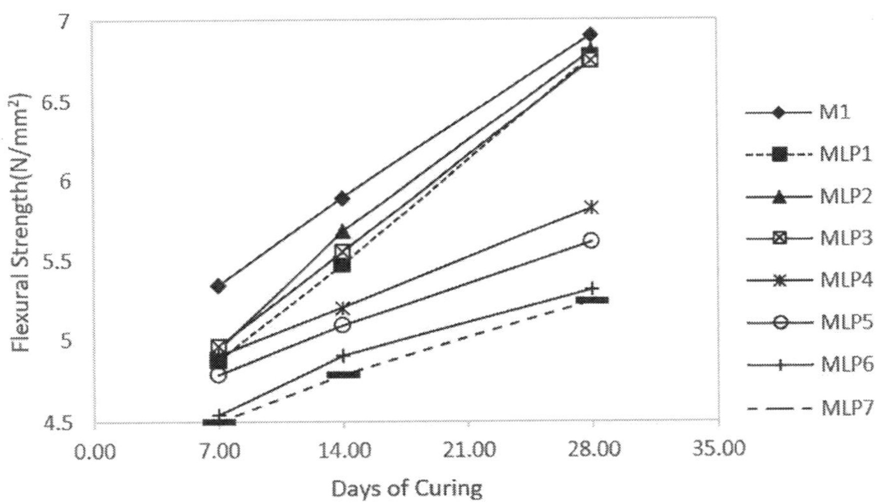

FIGURE 10.7 Comparison of flexural strength for multilayer plastic waste mix after 7, 14, and 28 days of curing (Bhogayata & Arora, 2017).

Discussion

1. The flexural strength of waste plastic mix concrete is decreased with the increase of the percentage of plastic waste in both cases, but the decrease in flexural strength is greater with waste plastic single layer compared to multilayer waste plastic.
2. The decrease in flexural strength is very little with the addition of 0.1% waste plastic single layer, and after that starts decreasing rapidly, and with 1% addition of waste plastic single layer, flexural strength decreases 28.7%.

This study has a positive impact on the environment, and the use of waste plastic in bitumen mix and concrete is eco-friendly.

1. Reduction in slump shows that it can be used only in required low degree workability, like in precast bricks, partition wall panels, and canal lining.
2. Density of concrete with multilayer waste plastic is more than the reference concrete up to 0.4% whereas, in the case of single-layer waste plastic, with the addition of 0.1%, there is a negligible change, but after that it starts decreasing. The density of concrete with multilayer waste plastic is more than single-layer waste plastic due to the higher specific gravity and density of the multilayer waste plastic.
3. Compressive strength decreases with the addition of waste plastic single layer, and with the addition of MLP waste, up to 0.3% compressive strength, increases, after that it starts decreasing.

REFERENCES

Bhogayata, A. C., & Arora, N. K. (2017). Fresh and strength properties of concrete reinforced with metalized plastic waste fibers. Construction and Building Materials, 146, 455–463.

Costa, De Brito, L. M., Silva, H. M., Oliveira, J. R., & Fernandes, S. R. (2012). Incorporation of waste plastic in asphalt binders to improve their performance in the pavement. International Journal of Pavement Research and Technology, 6(4), 457.

Ferreira, Luís, de Brito, Jorge, & Saikia, Nabajyoti. (2012). Influence of curing conditions on the mechanical performance of concrete containing recycled plastic aggregate. Construction and Building Materials, 36, 196–204.

Kandasamy, R., & Murugesan, R. (2011). Fibre reinforced concrete using domestic waste plastics as fibers. Journal of Engineering and Applied Sciences, 6(3), 75–82.

Kumar, A., Srivastava, V., & Kumar, R. (2014). Effect of waste polythene on compressive strength of concrete. Journal of Academia and Industrial Research (JAIR), 3(3), 152.

Merbouh, M. H., & Glaoui, B. (2015). Plastic waste in cement concrete, LDPE and PVC introduction effect. Advanced Materials Research, 1105, 325–330. https://doi.org/10.4028/www.scientific.net/amr.1105.325.

Raghatate Atul, M., & Saikia (2013). Use of plastic in a concrete to improve its properties. International Journal of Advanced Engineering Research and Studies, 1(3), 109–111.

Safi, Brahim, Saidi, Mohammed, Aboutaleb, Djamila, & Maallem, Madani. (2013). The use of plastic waste as fine aggregate in the self-compacting mortars: Effect on physical and mechanical properties. Construction and Building Materials, 43, 436–442.

11 Using Value Stream Mapping (VSM) as a Lean Production Tool

Streamlining the Production Process in a Lubricant and Degreaser Company

Erik Leonel Luciano, Rosinei Batista Ribeiro, Eliane Antônio Simões, Alexandre Formigoni, and Marcos Antonio Bessa

11.1 INTRODUCTION

Mosser (1977) states that, to eliminate waste, the value stream mapping (VSM) quality tool was developed through the Lean philosophy. Basically, through the VSM, it is possible to draw a "map" that shows how the flow of materials or information is, starting in the supply chain, passing through the company and ending with the customer, covering the entire path of the raw material transformation process. According to Marin-Garcia et al. (2021), it allows understanding the flow of a process through graphic elements and timelines. It maps the process as it is at the moment (current VSM) and, after extensive discussion, a new one is built (future VSM) (Morell-Santandreu et al., 2021). Although the principles of Lean thinking have been consolidated in several sectors, such as hospitals (Narayanamurthy et al., 2021), manufacturing (Godinho Filho et al., 2016) and services (Torri et al., 2021), the adoption of these principles in small- and medium-sized companies faces obstacles (Hu et al., 2015). Furthermore, the literature presents gaps regarding the application of Lean tools and the results that can be achieved (Belhadi et al., 2018). Therefore, given the gap in small- and medium-sized companies, this work will seek to answer the following research question: What opportunities for improvement could be found when applying VSM in a small company that manufactures lubricants and degreasers? To this end, the general objective was to identify possibilities for improvement and waste reduction in the production process of a lubricant and degreaser manufacturing industry through the application of the VSM tool. The methodology chosen was the case study through the execution of the following steps: definition of the

 DOI: 10.1201/9781003502647-13

team, selection of the product family, immersion, design of the current VSM and design of the rationalized future VSM. The research is characterized as qualitative and quantitative. As a result, the possibility of improvements was identified, including the combination of water insertion and raw material mixing operations, which could result in a reduction of up to 63.50% in T/C. In addition, several other improvement proposals were suggested through the 5W1H Kaizen action plan, without the need for large investments.

11.2 THEORETICAL REFERENCE

11.2.1 Lean Production

According to Deshmukh et al. (2022), the Lean manufacturing philosophy, introduced by Toyota, aims to minimize waste, specifying what is considered added value from the customer's point of view and separating activities that do not add value from those that do. For Kumar et al. (2022), the theory is built around two pillars: product quality, generated by a low-cost process and customer satisfaction, which is considered at the end of the production chain. Both approaches highlight the importance of efficiency and quality in production, aiming to meet customer expectations and reduce operating costs. Lead time comprises the time spent between ordering an item and it being available to the end customer, being one of the essential and important performance indicators for the management of manufacturing or service production processes (Gyulai et al., 2018; Lingitz et al., 2018; Kim et al., 2014). The word "takt" originates from the German language, meaning "rhythm" or "cadence", and refers to the regularity with which something is done, constituting a design parameter widely used in the Toyota Production System (Frandson et al., 2013). Milnitz (2018) explains that takt time can be understood as what drives the flow of materials in a production line or workstation. Chronoanalysis is fundamental to statistically measure and control the tasks to be carried out, allowing the calculation of the Standard Time (TP), which represents the time required to perform a task efficiently, considering normal working conditions (Jurandir & Reis, 2019). According to Mosser (1977), chronoanalysis consists of seven stages: summarize the operation, verify previous and subsequent operations to define beginning and end; subdivide operations into elements with defined beginning and end; record operator time with a stopwatch, clipboard and observation sheet; determine the number of timed cycles to obtain an arithmetic mean; Eevaluate the operator's rhythm considering effort and speed; define personal and fatigue tolerances for breaks and avoid fatigue; and determine standard time by dividing into normal time and tolerances. Tables 11.1 and 11.2 are used to calculate the determination of the number of cycles. Finally, Jardim (2014) and Okano et al. (2021) also explain that the skill and effort that the operator puts into his activities can be classified by monitoring the operator during the development of his tasks. The productive capacity, which, for Slack et al. (2009), is the maximum quantity of products and/or services that can be produced in a production unit, considering a period of time.

In Table 11.3, the classification concepts should be applied to help classify the range to be applied together with Table 11.4.

TABLE 11.1
Normal Distribution

Probability (%)	90	91	92	93	94	95
Z	1.65	1.70	1.75	1.81	1.88	1.96

Source: Adapted from Martins and Alt (2017).

TABLE 11.2
Coefficient to Calculate the Number of Timings

N	2	3	4	5	6	7	8	9	10
D2	1.128	1.693	2.059	2.326	2.534	2.704	2.847	2.97	3.078

Source: Adapted from Martins and Alt (2017).

TABLE 11.3
Efficiency Assessment Coefficients

Efficiency Calculation					
Ability (%)				**Effort (%)**	
120	A	Higher	120	A	Higher
115	B	Excellent	115	B	Excellent
110	C1	Good	110	C1	Good
105	C2	—	105	C2	—
100	D	Normal	100	D	Normal
95	E1	Regular	95	E1	Regular
90	E2	—	90	E2	—
85	F1	Weak	85	F1	Weak
80	F2	—	80	F2	—

Source: Adapted from Jardim (2014).

VSM is a powerful tool to identify waste and non-value-added activities in the production process of a specific product. By applying Lean production methods and industrial engineering, VSM is able to significantly improve the operational efficiency of the production system (Forno et al., 2014; Matt, 2014; Peralta et al., 2020). To perform the VSM, Mike and John (2012) propose a series of steps to be followed. The first is the choice of product family, which consists of selecting a group of products that go through similar processing steps. Next, it is necessary to design the current and future state of the process, based on the information collected on the shop floor. Finally, you should prepare an implementation plan that describes, on one page, how you want to get to the future state. According to Godoy et al. (2023), the Japanese methodology Kaizen originates from two words "Kai", which means

TABLE 11.4
Concepts for Classifying Ability and Effort

Ability
WEAK: Not adapted to work. Makes mistakes and has insecure movements.
REGULAR: You make fewer mistakes, and your movements are almost unsafe.
NORMAL: Satisfactory accuracy and consistent pace.
GOOD: He has confidence, a constant rhythm and rare excitement.
EXCELLENT: Precision in movements, without hesitations and errors.
HIGHER: Equal movements comparable to a machine.
Effort
WEAK: Lack of interest and inadequate methods.
REGULAR: Same as previous but with less intensity.
NORMAL: Constant work and satisfactory effort.
GOOD: Constant work and confidence, with little or no wasted time.
EXCELLENT: I work quickly and with precise movements.
HIGHER: Rhythms impossible to be maintained in a full day of work.

Source: Adapted from Jardim (2014).

change, and "Zen", which means for the better, that is, changing for the better, which enables continuous improvements in the organization. When applying Kaizen, tools and action plans can be used to put the desired improvements into practice (Burtet, 2022).

11.3 METHODOLOGY

This research is characterized as qualitative and quantitative. In terms of methodology, the case study was chosen, following the steps defined by Mike and John (2012), such as definition of the team, selection of the product family, immersion, design of the current VSM, design of the rationalized future VSM and, finally, the elaboration of a 5W1H Kaizen action plan with suggestions for improvements (Figure 11.1).

11.4 RESULTS AND DISCUSSION

11.4.1 Team Formation (First Stage)

Currently, the company has a total of 7 (seven) employees, including production, research and development laboratory, sales and administration. After the survey it was found that all employees should be included in the team.

11.4.2 Product Family Selection (Second Stage)

For the selection of the product family, it was identified that for October 2022, there was a production schedule of 31,070, which was the month of application of this work. It was highlighted that the demand for antifoam products corresponds to about

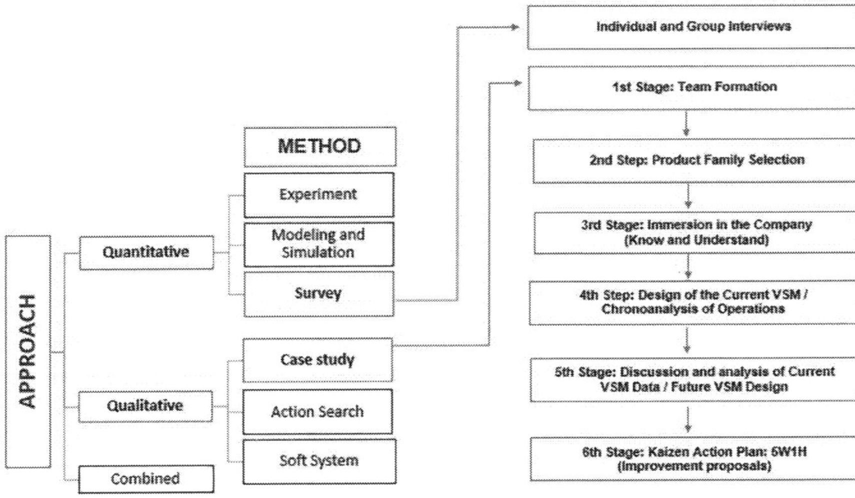

FIGURE 11.1 Work methodological flowchart.

Source: Authors (2023).

31.6% of the total produced, which is equivalent to 9 tons and 800 kg, and its production presents a great similarity with the way of manufacturing the other products, justifying its choice for the construction of the current VSM.

11.4.3 Immersion in the Company (Third Stage)

Packaging and Handling of Materials and Inventory of Raw Materials and Finished Products. Raw materials are received from suppliers in packaging in the form of 1000-liter containers, 70-kg barrels and fixed lid (TF) and removable lid (TR) drums, both 200 liters in carbon steel. A difference of the company in relation to drums of the removable cover type (TR), they are reused for bottling its manufactured products, washing and sanitizing the packaging beforehand.

11.4.4 Current VSM Design (Fourth Stage)

11.4.5 Identification of Improvement Points (Fifth Stage)

It was observed that to fill the exact amount of 190 liters per drum, one operator places the scale under the mixer while another places the drum on top; however, due to the elevation caused by the scale, the drum is very close to the outlet valve of the mixer product. It should also be noted that the operators perform this task in an uncomfortable and non-ergonomic position, requiring physical effort to lower it again. The base of the scale used still moves, increasing the risk of accidents in the work environment. Another operation that needs improvement is related to the

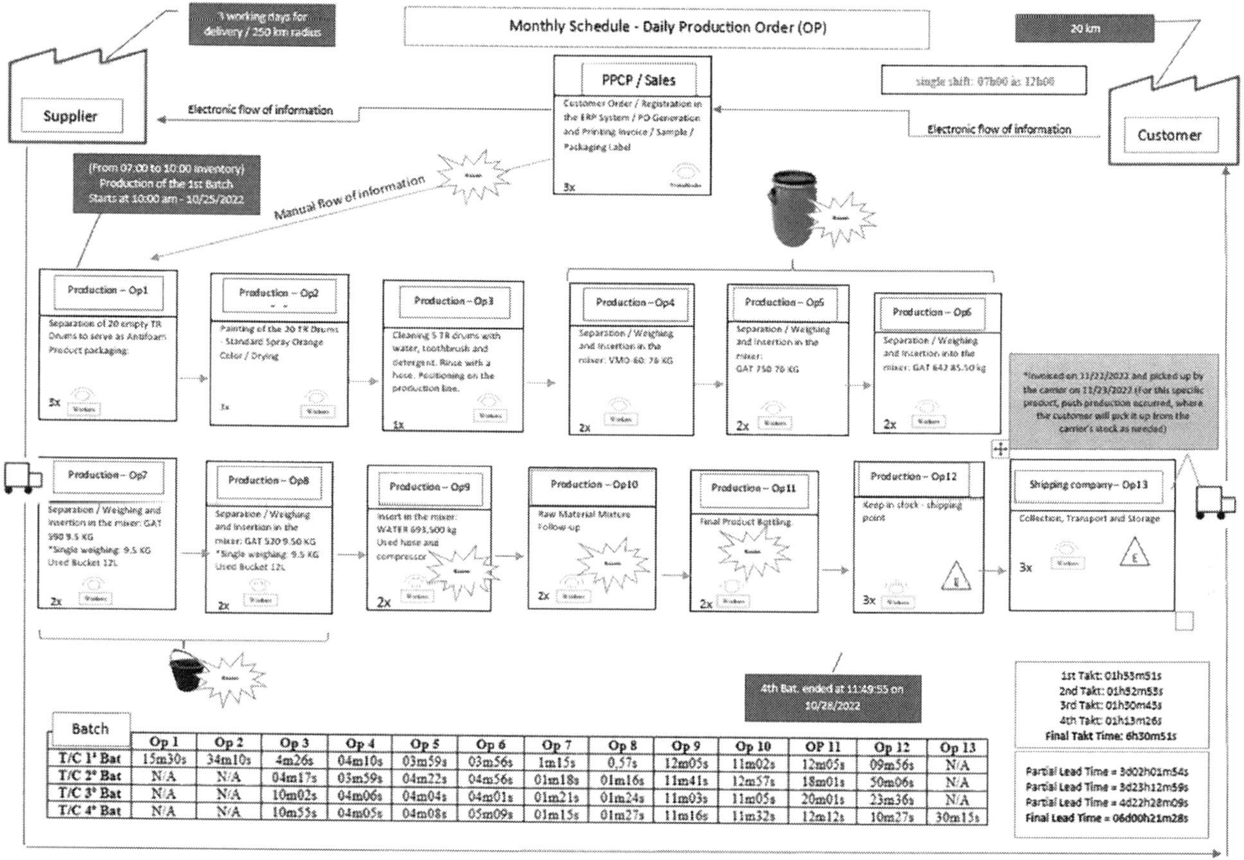

Batch	Op 1	Op 2	Op 3	Op 4	Op 5	Op 6	Op 7	Op 8	Op 9	Op 10	OP 11	Op 12	Op 13
T/C 1ª Bat	15m30s	34m10s	4m26s	04m10s	03m59s	03m56s	1m15s	0,57s	12m05s	11m02s	12m05s	09m56s	N/A
T/C 2ª Bat	N/A	N/A	04m17s	03m59s	04m22s	04m56s	01m18s	01m16s	11m41s	12m57s	18m01s	50m06s	N/A
T/C 3ª Bat	N/A	N/A	10m02s	04m06s	04m04s	04m01s	01m21s	01m24s	11m03s	11m05s	20m01s	23m36s	N/A
T/C 4ª Bat	N/A	N/A	10m55s	04m05s	04m08s	05m09s	01m15s	01m27s	11m16s	11m32s	12m12s	10m27s	30m15s

FIGURE 11.2 Drawing of the current VSM with the production synthesis of the four batches.

Source: Authors (2023).

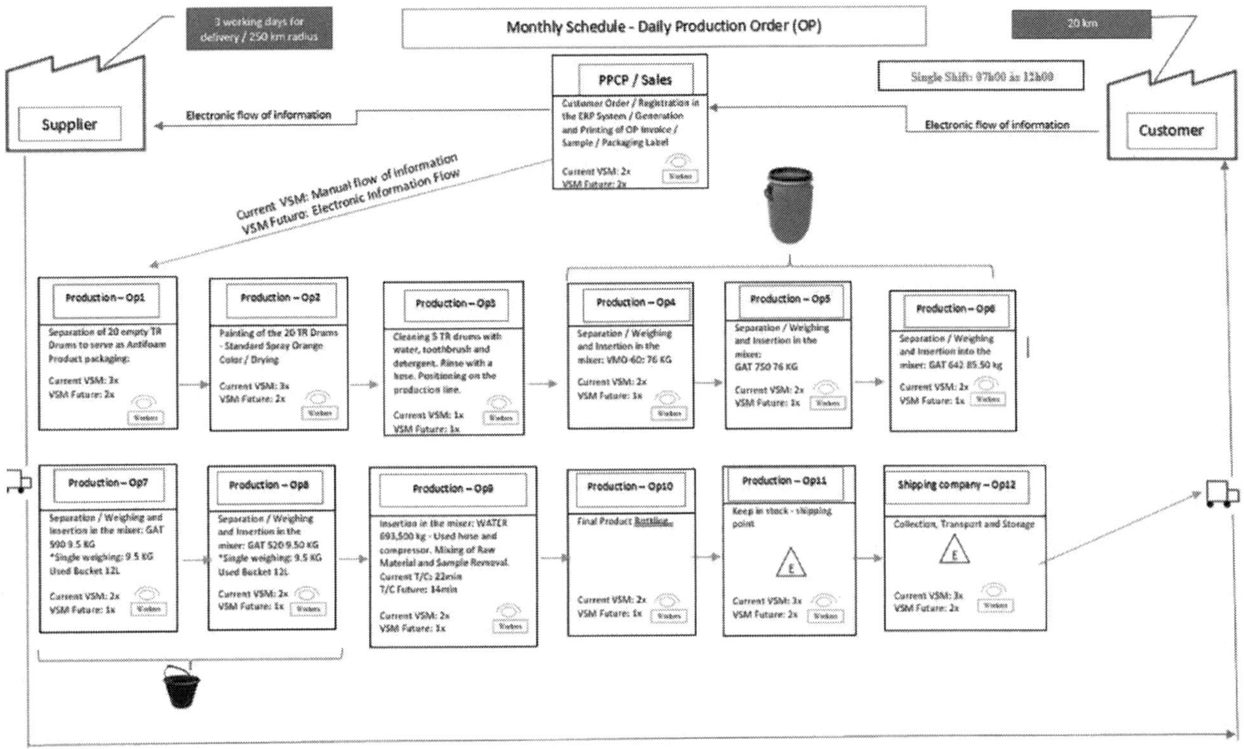

FIGURE 11.3 Future VSM design.

Source: Authors (2023).

insertion of the raw material into the mixer through drums with a handle, with a capacity of up to 50 liters.

11.4.5.1 VSM Future Design

The cycle time of the four batches was timed and calculated, however, to facilitate the understanding by the reader of the VSM Future was developed containing only the main operations to demonstrate which points can be optimized (Figure 11.3) and demonstrates that, in the development of the VSM Future design, improvements were focused on joining and optimizing some operations and on rebalancing the workforce on the production line. In all operations, with the exception of operations 09 and 10, which were merged, with the implementation of improvements suggested in the Kaizen plans, they needed to be quantified to map the difference in optimized execution time. Furthermore, some optimization suggestions will be further detailed in step 6, referring to Kaizen action plans: 5W1H.

11.4.6 Kaizen Action Plan: 5W1H—Improvement Proposals (Sixth Stage)

TABLE 11.5
Suggested Improvement Proposals—Kaizen 5H1W

Kaizen 1

(What): Optimization of the Production Process—Process No. 11 (Product filling in the TR drum)
(Why): In operation 11, it was observed that the elevation caused by the scale of the Welmy or Michalette models causes the drum to be very close to the mixer product outlet valve and the base of the scale in both models moves, not being an ergonomic operation that can lead to an accident at work.
(Where): Production line
(Who): Management
(When): Next 6 months
(How): Introduction of floor scales or implementation of pneumatic hoists with grippers adapted for drums or implementation of pneumatic hoists, gantry cranes and grippers adapted for handling.

Kaizen 2

(What): Optimization of the production process—merging operations No. 9 and 10
(Why): This will reduce the time of the two operations from 11 minutes each to 14 minutes together, enabling a reduction of approximately 63.5%, without changing the quality of the product.
(Where): Production line
(Who): Management
(When): Immediately
(How): Train employees so that when they start the water filling process, they turn on the mixer.

Kaizen 3

(What): Issuance of the production order (PO) in digital mode
(Why): Minimize printing and storage costs, as well as improve production efficiency, allowing employees to access necessary information more quickly and easily.
(Where): In the OP form

(*Continued*)

TABLE 11.5 (*Continued*)

(Who): To be noted by the operator

(When): Immediately

(How): Installation of a computer with a printer. Train employees and adapt so that the PO signature is done by the operator in digital format. Store documents electronically and use free data sharing tools such as Microsoft OneDrive, for example, to synchronize the desktops of the PPCP computer and the production line.

Kaizen 4

(What): Optimization of space in the manufacturing area

(Why): Little space for movement and storage of raw materials and finished products

(Where): At the company's headquarters

(Who): Management

(When): Next two years

(How): Acquire modular racks and easels for drum storage

11.5 CONCLUSIONS

Among the numerous opportunities for improvement found in the company with the application of the VSM tool, we can highlight possibility of optimization of operations 09 and 10 through their merger, resulting in a decrease of up to 63.50% TC; opportunity to rebalance the workforce across the entire production line with the implementation of several suggestions for improvements, which may lead to the optimization of labor in some operations by up to 50%; determination of the company's production capacity for the manufacture of antifoam: it is possible to produce a batch of the product every 1 hour and 30 minutes, resulting in a total daily production of 2850 kg of this product; possibility of optimizing storage space with the use of modular easels, for example, among others. As future work, it was verified that it is extremely important to carry out a more detailed ergonomic analysis of the operators who work directly in the production process through the RULA ergonomic analysis technique. This proactive approach is critical to managing health and safety in the workplace, thereby ensuring a better quality of life for workers and a safer, more productive work environment.

REFERENCES

Belhadi, A., Sha'ri, Y. B. M., Touriki, F. E., & El Fezazi, S. (2018). Lean production in SMEs: Literature review and reflection on future challenges. Journal of Industrial and Production Engineering, 35(6), 368–382. https://doi.org/10.1080/21681015.2018.1508081.

Burtet, A. (2022). Proposta de melhoria da logística através do uso de ferramentas da qualidade no centro de distribuição de Chapecó da Empresa Ex. https://rd.uffs.edu.br/handle/prefix/5663.

Deshmukh, M., Gangele, A., Gope, D. K., & Dewangan, S. (2022). Study and implementation of lean manufacturing strategies: A literature review. Materials Today: Proceedings, 62, 3309–3315. https://doi.org/10.1016/j.matpr.2022.02.155.

Forno, A. J. D., Pereira, F. A., Forcellini, F. A., & Kipper, L. M. (2014). Value stream mapping: A study about the problems and challenges found in the literature from the past 15 years about application of lean tools. The International Journal of Advanced Manufacturing Technology, 72, 779–790. https://link.springer.com/article/10.1007/s00170-014-5712-z.

Frandson, A., Berghede, K., & Tommelein, I. D. (2013, August). Takt time planning for construction of exterior cladding. In 21st Annual Conference of the International Group for Lean Construction. Fortaleza, Brazil (pp. 527–536). https://iglcstorage.blob.core.windows.net/papers/attachment-3465d25c-5283-47dd-9d3b-106ed6416d70.pdf.

Godinho Filho, M., Ganga, G. M. D., & Gunasekaran, A. (2016). Lean manufacturing in Brazilian small and medium enterprises: Implementation and effect on performance. International Journal of Production Research, 54(24), 7523–7545. http://dx.doi.org/10.1080/00207543.2016.1201606.

Godoy, R. F., Munhoz, I. P., de Oliveira, G. G., Rosa, M. T. D. M. G., de Gouvêa, M. T., & Akkari, A. C. S. (2023). Operacionalização da metodologia Kaizen a fim de otimizar o desempenho organizacional: Um enfoque na gestão de processos em uma indústria eletrônica. Revista de Gestão e Secretariado, 14(2), 1832–1854. https://revistagesec.org.br/secretariado/article/view/1665.

Gyulai, D., Pfeiffer, A., Nick, G., Gallina, V., Sihn, W., & Monostori, L. (2018). Lead time prediction in a flow-shop environment with analytical and machine learning approaches. IFAC-PapersOnLine, 51(11), 1029–1034. https://doi.org/10.1016/j.ifacol.2018.08.472.

Hu, Q., Mason, R., Williams, S. J., & Found, P. (2015). Lean implementation within SMEs: A literature review. Journal of Manufacturing Technology Management, 26(7), 980–1012. https://doi.org/10.1108/JMTM-02-2014-0013.

Jardim, M. L. C. (2014). Manual de avaliação e desempenho humano nas atividades industriais. Rio de Janeiro: Editora Interciência.

Jurandir., P., & Reis, G. A. (2019). Engenharia de métodos: aplicação e economia na produção. Atlas.

Kim, S. H., Kim, J. W., & Lee, Y. H. (2014). Simulation-based optimal production planning model using dynamic lead time estimation. The International Journal of Advanced Manufacturing Technology, 75, 1381–1391. https://doi.org/10.1007/s00170-014-6209-5.

Kumar, N., Hasan, S. S., Srivastava, K., Akhtar, R., Yadav, R. K., & Choubey, V. K. (2022). Lean manufacturing techniques and its implementation: A review. Materials Today: Proceedings, 64(3), 1188–1192. https://doi.org/10.1016/j.matpr.2022.03.481.

Lingitz, L., Gallina, V., Ansari, F., Gyulai, D., Pfeiffer, A., Sihn, W., & Monostori, L. (2018). Lead time prediction using machine learning algorithms: A case study by a semiconductor manufacturer. Procedia CIRP, 72, 1051–1056. https://www.sciencedirect.com/science/article/pii/S2212827118303056.

Marin-Garcia, J. A., Vidal-Carreras, P. I., & Garcia-Sabater J. J. (2021). The role of value stream mapping in healthcare services: A scoping review. International Journal of Environmental Research and Public Health, 18(3), 951. https://doi.org/10.3390/ijerph18030951.

Martins, P. G., & Alt, P. R. C. (2017). Administração de materiais e recursos patrimoniais. Saraiva Educação SA.

Matt, D. T. (2014). Adaptation of the value stream mapping approach to the design of lean engineer-to-order production systems: A case study. Journal of Manufacturing Technology Management, 25(3), 334–350. https://doi.org/10.1108/JMTM-05-2012-0054.

Mike, R., & John, S. (2012). Aprendendo a enxergar: mapeando o fluxo de valor para agregar valor e eliminar desperdícios. São Paulo: Lean Institute Brasil.

Milnitz, D. (2018). Tempos e Métodos Aplicados à Produção. Indaial: Uniasselvi.

Morell-Santandreu, O., Santandreu-Mascarell, C., & Garcia-Sabater, J. J. (2021). A model for the implementation of lean improvements in healthcare environments as applied in a primary care center. International Journal of Environmental Research and Public Health, 18(6), 2876. https://doi.org/10.3390/ijerph18062876.

Mosser, B. R. (1977). Estudos de tempos e movimentos com fins de racionalização do trabalho. 13. ed. São Paulo: Edgard Blücher.

Narayanamurthy, G., Gurumurthy, A., & Lankayil, A. A. (2021). Experience of implementing lean thinking in an Indian healthcare institution. International Journal of Lean Six Sigma, 12(1), 23–60. https://doi.org/10.1108/IJLSS-10-2016-0062.

Okano, M. T., dos Santos, H. D. C. L., Ursini, E. L., Honorato, W., & Ribeiro, R. B. (2021, January). Digital transformation as an agent of change in a pharmaceutical industry from the perspective of dynamic capabilities. In 2021 IEEE 11th Annual Computing and Communication Workshop and Conference (CCWC) (pp. 0896–0902). IEEE. 10.1109/CCWC51732.2021.9375838

Peralta, C. B. D. L., Echeveste, M. E., Lermen, F. H., Marcon, A., & Tortorella, G. (2020). A framework proposition to identify customer value through lean practices. Journal of Manufacturing Technology Management, 31(4), 725–747. https://doi.org/10.1108/JMTM-06-2019-0209.

Slack, N., Chambers, S., & Johnston, R. (2009). Administração da produção (Vol. 2). São Paulo: Atlas.

Torri, M., Kundu, K., Frecassetti, S., & Rossini, M. (2021). Implementation of lean in IT SME company: An Italian case. International Journal of Lean Six Sigma, 12(5), 944–972. https://doi.org/10.1108/IJLSS-05-2020-0067.

12 Study of Design Requirements for the Design of Bakelite Materials for Smart Manufacturing Applications

Caio Fernando da Silva,
Rosinei Batista Ribeiro, Marco Aurélio Feriotti,
Alexandre Formigoni, Marcelino Rivas Santana,
Eliane Antônio Simões, and
Karolyne Borges de Oliveira

12.1 INTRODUCTION

The kitchen is the main room in a home where food is prepared for the family, which makes it important for any culture. It has been an integral part of human life for thousands of years, and its evolution reflects our changing needs and values over time (Franco, 2001). In addition, Da Silva Azevedo (2018) points out that there is currently a wide variety of utensils in the kitchen, from a simple vegetable peeler to more complex objects such as planetary mixers and the like. These objects have different models but the same function, which, by using varied materials, become different versions of the same product with unique properties that offer characteristics and advantages, enabling or facilitating specific food preparation processes. In this context, contact was made with the company Baquellites®, which produces and develops accessories for kitchen utensils using thermoplastic transformation material, in this case, bakelite, which has a specific demand for the development of panhandles that have specific technical and regulatory requirements, to guarantee an aesthetic connection that allows it to be paired with the handle developed by Baquellites. As such, for this product to be developed correctly, so that it is in line with the markets and companies' requirements, a list of prerequisites must be met. The project specification determines the specific objectives for the new product and will be those incorporated

DOI: 10.1201/9781003502647-14

into the product and offered to the consumer. It is therefore especially important that the project specification is done well so that the new product is developed correctly (Baxter, 2018). This work aims to explore the design demands that contextualize the designing and manufacturing of bakelite kitchen appliances by exploring the relationship between the product, manufacturing, and context, to present valuable insights that may be used to positively influence the future the designing of new products. This project derives from the study of a branch of metal hybrid additive manufacturing (MAHM) in the reuse of injection molds to produce kitchen appliances, such as handles, in bakelite. As Iqbal and Suzianti (2021) state, this project is about technological development and innovative extension with the interaction of demands and competencies in full constructive collaboration between the university, the company, and the government.

12.2 THEORETICAL REVIEW

12.2.1 Product Development

According to the definitions and descriptions presented by Otto (2003), Baxter (2018), and Rosato and Rosato (2012), the term "product design" can be described as the process of creating and developing a new product that meets the needs and expectations of the target market while considering the final user as its main focus. It is a multidisciplinary approach that integrates elements of engineering, marketing, ergonomics, and psychology, among other fields. Product design can have a significant impact on the success of a product on the market, as it influences the user's experience and the perceived value of the product. According to Kabuchi and Ribeiro (2021), "Design is not just what it looks and feels like. Design is how it works." This statement highlights the importance of functional design in product development. A product must not only look and feel good, but it must also perform well and meet the user's needs. In addition to functional design, ergonomic design is crucial in product development. As stated by Don Norman (1988), "Good design is much harder to notice than bad design, partly because good designs fit our needs so well that the design is invisible." Even while considering the attention to the marketing and industrial aspects that permeate product conception, the designer's focus when designing products is on the user, who not only acquires the product but also uses it.

The activity of developing a new product requires research, careful planning, meticulous control, and the use of systematic methods. Systematic design methods require an interdisciplinary approach, encompassing engineering methods, ergonomics, marketing, and the application of knowledge about aesthetics and style. The need for innovation requires the coming together of social sciences, technology, and applied art (Baxter, 2018; Ribeiro et al., 2013).

Product design plays a crucial role in integrating various disciplines to create functional, ergonomic, and socially responsible products. Research (Instituto Nacional de Metrologia, Normalização e Qualidade Industrial—INMETRO, 2012) highlights the importance of the new product development (NPD) process for small- and medium-sized enterprises (SMEs). In this process, open innovation emerges as a valuable strategy for SMEs to overcome resource constraints and market uncertainties. Tools such as simulation, crowdsourcing, and the implementation of quality functions offer support during the design phase. In addition, Klun et al. (2022) highlight sustainability as a key factor for market success in product design. They emphasize the eco design integration method for SMEs as a tool that facilitates environmental considerations in the NPD process (Le Pochat et al., 2017). Collaboration with stakeholders, as pointed out by Norman (1988), is also crucial. Crowdsourcing has been shown to be effective in generating ideas, improving product quality, and reducing time and cost.

12.2.2 Design and Selection of Materials

Materials have been with us forever. Their association with culture dates back to the Stone Age, the Bronze Age, and the Iron Age, among others (Callister et al., 2007). To this day, materials embody beliefs, attitudes, and values that can be expressed in products with an artistic bias (Domingos & Ribeiro, 2015). As noted by Kelley and Kelley (2013), the importance of selecting materials is also because materials can have a significant impact on the environment and health.

It is materials that give substance to everything we see and touch. *Homo sapiens* is recognized for its ability to design and produce objects from materials and to see more in an object than its appearance, which can have meaning, arouse associations, or be signs of more abstract ideas. Object designs, both symbolic and utilitarian, precede any recorded language, and give us the earliest evidence of a cultural society and symbolic reasoning (Ashby & Johnson, 2013).

It is through materials that we perceive the world, be it through appearance, touch or, cording (Ashby & Johnson, 2013), even shape, and it is essential to pay due attention to this aspect, because this way it is possible to have greater control over the user's perception of the product, as well as enabling new ways of controlling the cost of the product, either by saving on quantity through better knowledge of mechanical properties or by using better-selected materials. Regarding the material used in the product in question, it is worth mentioning that bakelite, as the first fully synthetic plastic that revolutionized the manufacture of various consumer products, is still widely used in the automotive industry due to its lightness. However, it is essential to note that bakelite is fragile and can break down into microplastics under high mechanical pressure or abrasion, thus impacting wildlife and releasing harmful chemicals into the environment (American Chemical Society, n.d.). It is therefore crucial to consider environmental and health effects when choosing materials, even if they are of historical importance, as with bakelite.

12.3 METHODOLOGY

To develop this project, an integration of different methodologies was employed. This led to the convergence of the methodological stages proposed by Baxter, Munari, and Frascara (Figure 12.1). The application and use of methodologies aim to simplify and improve the design process, making it easier for designers to conceive, develop, implement, and operate their projects effectively and efficiently. The technological development and innovative extension project were carried out in a company located in the East Zone of São Paulo—SP, in the sector of transformation of polymeric materials, such as bakelite.

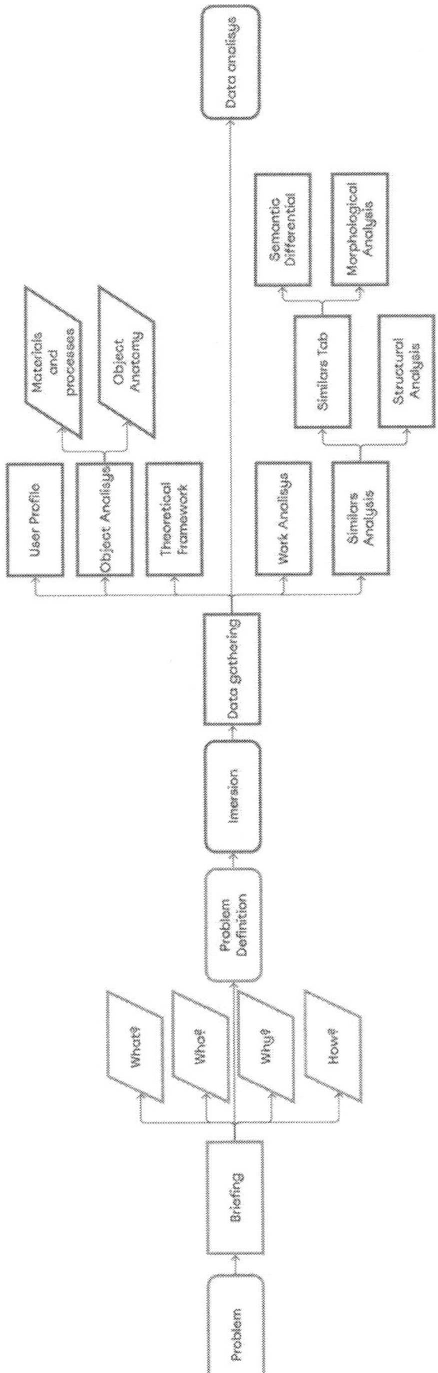

FIGURE 12.1 Flowchart showing the data collection methodology and its stages.

The integration of the Baxter, Munari, and Frascara methodologies provides a comprehensive and comprehensive framework for the development of design projects, streamlining the process and reducing the complexity involved, resulting in a streamlined and organized development process that incorporates innovative tools to help identify an optimal solution to the problem of this project (Feriotti et al., 2021).

12.4 RESULTS AND DISCUSSION

The integration of the converged methodology has provided support that can be used while designing products, for instance, dealing with the complexity of integrating technologies such as metal hybrid additive manufacturing with the improvement of mechanical properties, reduction of the coefficient of friction, and wear in the relationship between the product and the mold, as well as the establishment of an environment that supports future analysis and studies directly with the user.

12.4.1 Defining Demand

As part of the Baquellites original product development process, a handle was developed to make "French" frying pans, which are characterized by the elevation of the handle and the handle in relation to the height of the pan, shown in Figure 12.2.

However, even though the company's current product catalog has handles that can be attached to the frying pan together with the new handle, these previously produced handles do not meet the aesthetic requirements that would make them capable of pairing, that is, being introduced together with the handle in such a way that they

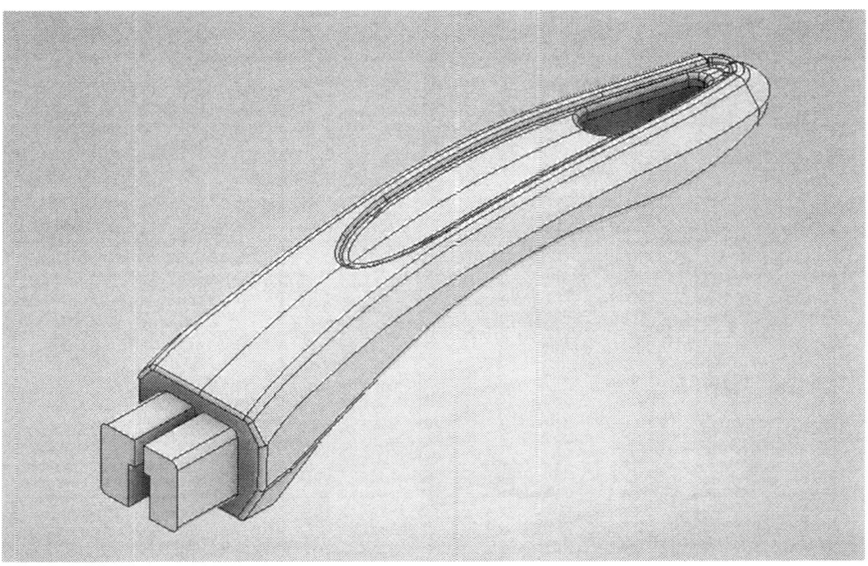

FIGURE 12.2 Cable developed by Baquellites.

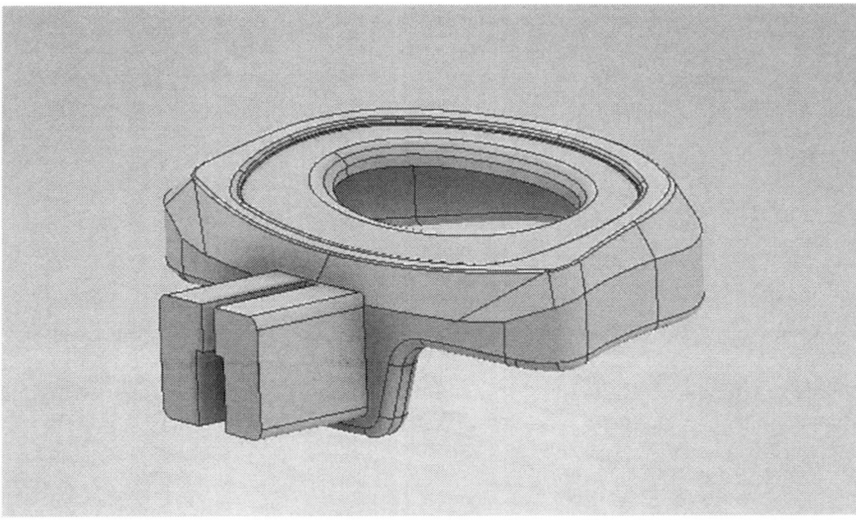

(A)

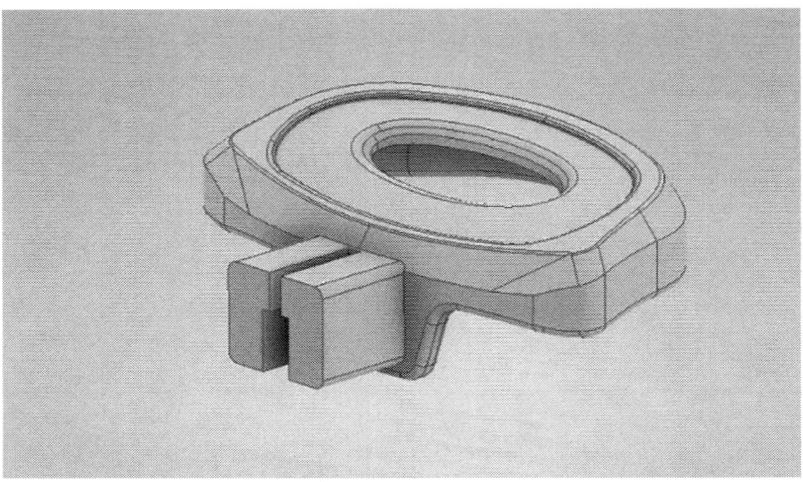

(B)

FIGURE 12.3(A) AND (B) 3D modeling of the rejected handles.

have a design, technical, ergonomic, and aesthetic relationship judged to be adequate by the company (Figures 12.3 (a) and (b)).

12.4.2 Normative Specifications

There are technical regulatory requirements that the product must meet before it can be marketed on a large scale (see Table 12.1). In Brazil, these standards are regulated by Inmetro's (National Institute of Metrology, Quality and Technology)

TABLE 12.1
INMETRO's Technical Requirements for Straps, Handles, and Pommels

Requirements	Specification/Target
Torsion resistance	They must withstand a torsional moment of 5 N.m ± 0.5, without fractures and/or cracks visible to the naked eye.
Fatigue	They must withstand 1500 test cycles in accordance with section 9 of technical standard ABNT NBR 1487, 1 (one) cycle per minute, without showing fractures and/or cracks visible to the naked eye.
Heat propagation	The maximum temperature for handles, straps, or lozenges applied to the body of the utensil must be up to 50°C and 60°C for handles, straps, or lozenges applied to the lid of the utensil.
	It must not be possible to reach with a 14 mm diameter spherical probe any part of the metal contained within the cable, handle, or pommel whose temperature exceeds 55°C.
Burn resistance	Cables and handles must not melt or spread flames when a flame is applied to any part of their extension.
Heat resistance	The handles, grips, and pommels, when used for the purpose of removing utensils that will be used in ovens, when placed in an oven with an internal temperature of 180°C ± 2 for a period of 1 (one) hour, and then cooled to room temperature for three hours, must not show any damage or distortion either to themselves or to the fixing systems of the body or lid of the utensil.

technical quality regulation for metal cookware, based on Standard NBR14876: Metal Household Utensils—handles, pommels, and fastening systems, 2016 (Associação Brasileira de Normas Técnicas, 2016).

12.4.3 Operational Demands

For the product to be produced by Baquellites, it is essential to meet operational requirements, such as precise sizing and standard fittings. These criteria guarantee the compatibility and correct assembly of the product, ensuring its efficiency and functionality during production and use. These requirements include the following specifications:

Maximum height of 80 mm; maximum width 100 mm; maximum depth of 90 mm; it must contain the standardized fitting so that it can be coupled to the structure that attaches it to the pan (Figure 12.4).

Bakelite products are produced by injection molding, in which the amorphous mass of bakelite is inserted into the mold cavity, where curing takes place, and the material solidifies, taking on the contour of the mold [23]. The insertion site, where the material was injected into the mold, does not take the molded shape, resulting in the formation of a cylindrical prominence categorized as a "burr," which is removed shortly afterward. This process leaves a small mark on the spot (Figure 12.5).

To meet customer demands regarding the aesthetic details of the products, Baquellites has introduced a new production process in which the insertion point is hidden. Although this solves the previous issue, it has resulted in a new challenge. The new configuration directs the injection perpendicular to the mold wall, leading

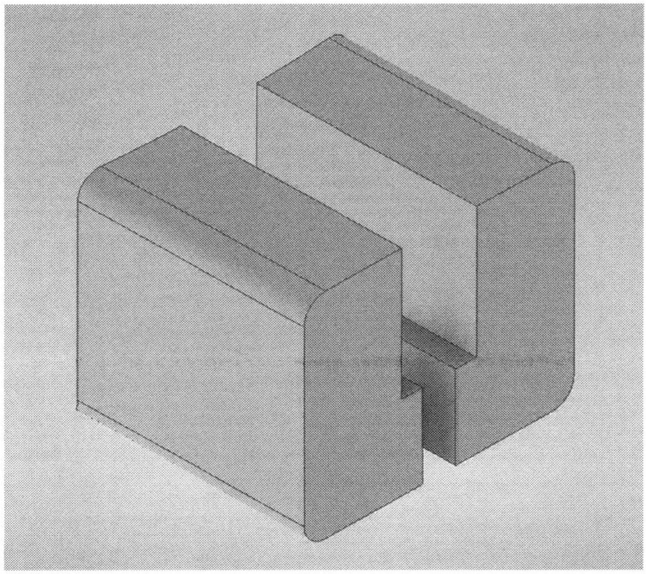

FIGURE 12.4 3D model of the area where the handle fits into the support.

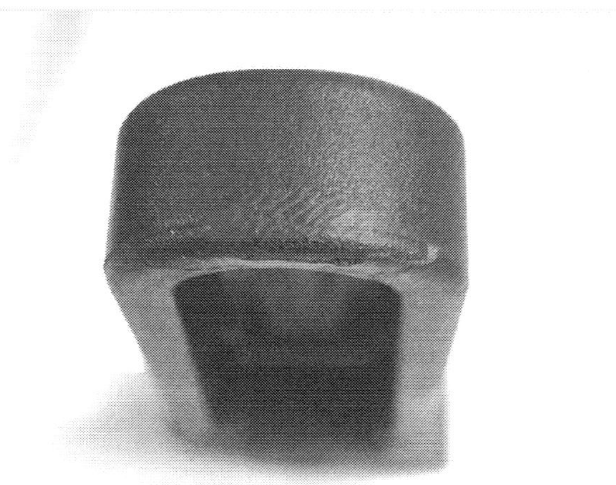

FIGURE 12.5 Detail of the mark left at the injection insertion point.

to friction-induced abrasion that reduces the useful life of the mold. This abrasion can damage the surface of the mold, removing the texture designed for the final product and altering its shape (Figures 12.6 (a) and (b)).

Authors claim that there is work being done to solve this problem using additive manufacturing, especially MAHM, which has the potential to play a key role in the

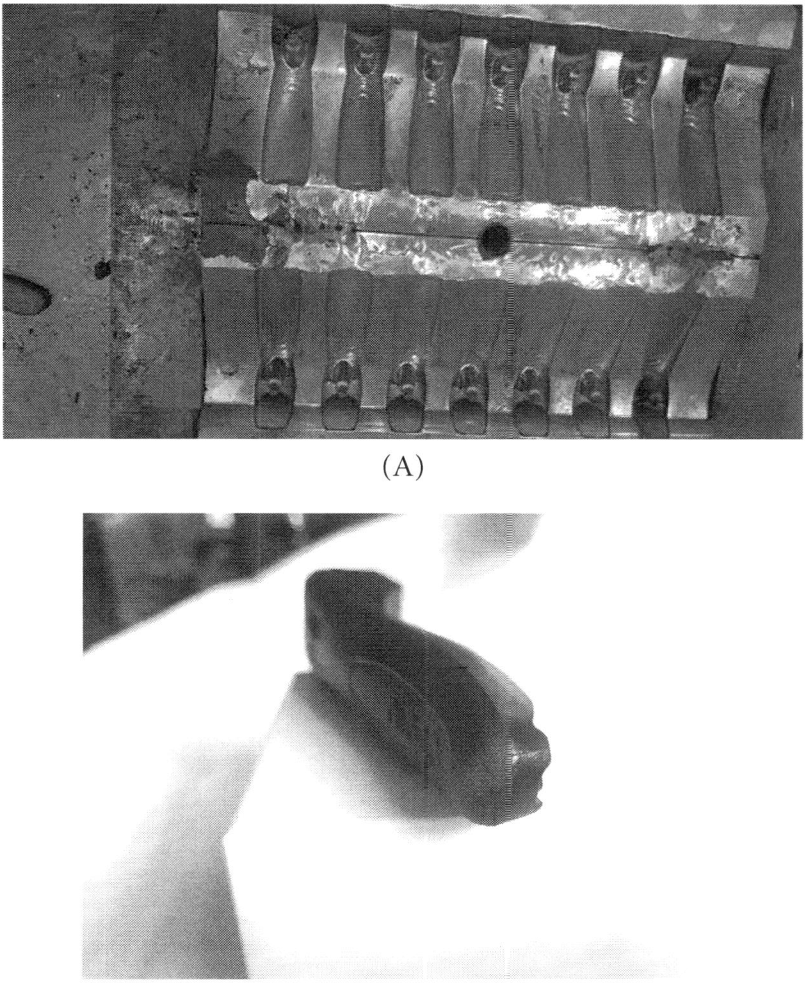

(A)

(B)

FIGURE 12.6(A) AND (B). Inner surface of the mold damaged by injection (b) Bakelite handle with surface damaged by abrasion in the mold.

product design process. Arrizubieta et al. (2020) have analyzed the environmental implications of using metal powder in additive manufacturing, Basinger et al. (2018) developing a modular computer-aided process planning system and Almeida (2021) identifying barriers and proposing a roadmap for the adoption of additive manufacturing in companies. In addition, Norman (2004) highlights the impacts of additive manufacturing in the manufacture of thermoplastic injection molds, and the systematic literature review by Feriotti and Formigoni (2022) addresses the applications, characteristics, and mechanical properties of different types of steels in hybrid metal additive manufacturing. These studies highlight the promising role of hybrid additive manufacturing of metals in improving the product design process and in the search for more efficient and sustainable solutions and influence design due to Baquellites's intention to apply such processes.

12.5 FINAL CONSIDERATIONS

This research aims to highlight the relationships between the product, manufacturing, and the context in which it originates, in the integration between the company, university, and government, now demand and competence, given the nature of applied research. It was evident that the integration and convergence of the different methodologies in their stages through the modular design that facilitates the various types of work, the development of the product within the normative parameters, in accordance with standard NBR 14876-16, the implementation in the construction of the molds in their recovery through the use of new techniques of MAHM and in the operationalization with the study of the feasibility of cost and use of the mold in full production on a daily basis. In turn, the main objective of the research was to delve into the intricate relationships between the product, its production, and the context in which it is conceived. The study sought to gather data to provide a better understanding of the factors that shape and influence the product. This information constitutes an essential knowledge base for reference and validation of ideas and future decisions. Given this, the objective of this research was achieved by understanding the product's design context, analyzing its design demands, and gathering a relevant amount of data that can be employed as a valuable resource for the following development, providing a basis for future research and decision-making, and, ultimately, helping to ensure the product's long-term success.

REFERENCES

Almeida, J. F. D. (2021). Adoção de manufatura aditiva (MA) para metais em empresas: Identificação de barreiras e proposta de roteiro para implementação (Doctoral dissertation, Universidade de São Paulo). https://doi.org/10.11606/D.18.2021.tde-23082021-180414

American Chemical Society. (n.d.). Bakelite® first synthetic plastic—National Historic Chemical Landmark. https://www.acs.org/education/whatischemistry/landmarks/bakelite.html

Arrizubieta, J. I., Ukar, O., Ostolaza, M., & Mugica, A. (2020). Study of the environmental implications of using metal powder in additive manufacturing and its handling. *Metals*, *10*(2), 261. https://doi.org/10.3390/met10020261

Ashby, M., & Johnson, K. (2013). *Materiais e design: A arte e ciência da seleção de materiais no projeto do produto*. Elsevier Brasil.

Associação Brasileira de Normas Técnicas. (2016). *NBR 14876: Utensílios domésticos metálicos—Alças, cabos, poméis e sistemas de fixação*. ABNT.

Basinger, K. L., Keough, C. B., Webster, C. E., Wysk, R. A., Martin, T. M., & Harrysson, O. L. (2018). Development of a modular computer-aided process planning (CAPP) system for additive-subtractive hybrid manufacturing of pockets, holes, and flat surfaces. *The International Journal of Advanced Manufacturing Technology*, *96*, 2407–2420. https://doi.org/10.1007/s00170-018-1674-x

Baxter, M. (2018). *Product design*. CRC Press.

Callister, W. D., Rethwisch, D. G., Blicblau, A., Bruggeman, K., Cortie, M., Long, J., & Mitchell, R. (2007). *Materials science and engineering: An introduction* (7th ed.). John Wiley & Sons.

Da Silva Azevedo, A. S. (2018). Design de produto e comunicação na F. Carvalho, Lda (Doctoral dissertation, Universidade da Beira Interior, Portugal). http://hdl.handle.net/10400.6/9685

Domingos, B. S. M., & Ribeiro, R. B. (2015). Geração de renda informal e desenvolvimento econômico: Tecnologias sociais como uma alternativa à precarização. *Quanta: Comunicação e Cultura, 1*(1), 78–83. https://www.aedb.br/publicacoes/index.php/comunicacao/article/viewFile/11/10

Ferlotti, M. A., & Formigoni, A. (2022). Manufatura aditiva híbrida de metais (MAHM), aplicações e características, tipos de aços e propriedades mecânicas: Uma revisão sistemática da literatura. In *XVII Simpósio dos Programas de Mestrado Profissional, São Paulo.*

Ferlotti, M. A., de Medeiros Marcelino, D., Pohlmann, M. N., Neto, J. M., & Rosa, J. L. (2021). Aplicações da manufatura aditiva e impressão 3d na fabricação de moldes para injeção de termoplásticos. *Brazilian Journal of Production Engineering, 7*(3), 199–218.

Franco, A. (2001). De caçador a gourmet: Uma história da gastronomia. Thesaurus Editora.

Instituto Nacional de Metrologia, Normalização e Qualidade Industrial—INMETRO. (2012). *Regulamento técnico da qualidade para panelas metálicas.*

Iqbal, M., & Suzianti, A. (2021). New product development process design for small and medium enterprises: A systematic literature review from the perspective of open innovation. *Journal of Open Innovation: Technology, Market, and Complexity, 7*(2), 153. https://doi.org/10.3390/joitmc7020153

Kabuchi, T. A. F., & Ribeiro, R. B. (2021). Manufatura aditiva em materiais metálicos, seus processos e aplicações—Um ensaio teórico. *Revista de Administração do UNIFATEA, 22.*

Kelley, T., & Kelley, D. (2013). *Creative confidence: Unleashing the creative potential within us all.* Currency.

Klun, B., Rozman, U., Ogrizek, M., & Kalčíková, G. (2022). The first plastic produced, but the latest studied in microplastics research: The assessment of leaching, ecotoxicity and bioadhesion of Bakelite microplastics. *Environmental Pollution, 307*, 119454. https://doi.org/10.1016/j.envpol.2022.119454

Le Pochat, S., Bertoluci, G., & Froelich, D. (2007). Integrating ecodesign by conducting changes in SMEs. *Journal of Cleaner Production, 15*(7), 671–680. https://doi.org/10.1016/j.jclepro.2006.01.004

Norman, D. A. (1988). *The psychology of everyday things.* Basic Books.

Norman, D. A. (2004). *Why we love (or hate) everyday things.* Perseus Books Group.

Otto, K. N. (2003). *Product design: Techniques in reverse engineering and new product development.* 清华大学出版社有限公司.

Ribeiro, R. B., Silva, J. W. J., Hein, L. R. O., Pereira, M. C., Codaro, E. N., & Matias, N. T. (2013). Morphology characterisation of pitting corrosion on sensitized austenitic stainless steel by digital image analysis. *International Scholarly Research Notices, 2013.* https://doi.org/10.1155/2013/905942

Rosato, D. V., & Rosato, M. G. (2012). *Injection molding handbook.* Springer Science & Business Media.

Index

A

abrasion and erosion in butterfly valves, 85–87, 91–95
additive and subtractive manufacturing integration, 3–9
additive manufacturing (AM) 3–18, 21–37, 59, 187–188
aerospace industry, 4, 12, 17, 66, 73, 74, 85
AI algorithms, 71
AM equipment, 11
AM equipment advances, 7, 11, 12
AM software, 10, 14, 15
AM software, specific software (ESPRIT CAM, Siemens NX) 14, 15, 18
AM training, 16
amorphous silica in sugarcane bagasse ash (SCBA), 99–109
annular fins in heat sink design, 148–152
application processes of metal additive manufacturing 5
applications of HMAM, 4, 16, 17, 18
applications of I4o technologies, *39*
artificial intelligence (AI), 60
artificial intelligence (AI) in smart manufacturing, 60–62, 63–73
artificial neural network (ANN), 86
augmented reality, 24, 58
automation, 60, 68–70, 78, 79
automotive industry applications, 13, 71, 181

B

bagasse ash concrete, 99–109
Bakelite material design and mold manufacturing, 179–188
benefits of HMAM, 5
big data, 59, 60, 65, 69, 114
binder jetting, 24
bituminous asphalt with plastic waste 154–167
blockchain, 75
business, 127, 128
butterfly valve erosion modeling, 85–90

C

CAM software, 10, 15, 18
car assembly robot, *184*
case studies in smart welding and casting, 71–74
challenges and limitations in traditional manufacturing, 4
challenges in HMAM, 3, 4, 10, 16, 18
challenges, 3, 12
cloud computing, 59, 65, 69, 78
cloud platforms in manufacturing, 64
CoCr (Cobalt Chrome), 8, 9
comparison of additive manufacturing process 68
computational fluid dynamics (CFD) in butterfly valve analysis, 85–90
computer-aided design (CAD), 25
concepts of intelligent welding, 65
concrete with plastic waste 154–167
concrete with sugarcane bagasse ash (SCBA), 99–109
conjugate gradient algorithm, 150
cooling system heat sink design, 142–152
crowd sourcing in product development, 180–183
cyber-physical systems (CPS), 59, 60, 64
cybersecurity, 72, 75

D

DED (direct energy deposition), 4, 5, 6, 10, 24
dentistry, 7
design science research (DSR) methodology, 126–134
differences between ESPRIT CAM and Siemens NX software, 15
diffuser and crushing mill in sugarcane juice extraction, 113–124
DPM and DEM models in CFD, 85–89
durability and mechanical properties of SCBA concrete, 99–105

E

electrical discharge machining (EDM), 33–35, 40
electrical resistivity of concrete with SCBA, 107–108
electronics industry, 17
energy efficiency, 127, 135, 136

energy generation of HDPE tiles with graphene, 126–139
energy industry, 17
environmental impacts of solar panels and graphene tiles 132–134
ergonomics in production process optimization, 169–176
erosion rate plot comparison, *94*
erosion wear in butterfly valves, 85–95
ESPRIT CAM, 14
ESPRIT CAM software, 14, 15
estimated global growth of the hybrid additive manufacturing, *14*

F

FANUC arc mate IoT-enabled robotic welding machine *64*
fatigue and wear in Bakelite mold production, 185–188
ferrofluid–based passive cooling systems 142–152
final considerations, 16–18
flexural strength of concrete with plastic waste, 166–167
flow characteristics and pressure in butterfly valves, 85–92
flowchart of RPA tool, *207*
frugal innovation and social technology concepts, 126–134

G

global patent analysis for sugarcane juice extraction, 113–124
graphene properties and application in HDPE roof tiles, 129–130
green technologies in smart manufacturing, 60–66
grinding process in sugarcane juice extraction, 113–124

H

heat sink design parameters and fin optimization, 148–151
HMAM (hybrid metal additive manufacturing) 3–18
human-centric approach 42, 53, 101
human–machine collaboration, 76
human–robot collaboration in smart manufacturing, 76–77
hybrid manufacturing machines, 7, 15, 16, 17
hybrid manufacturing software 10, 14, 15
hybrid metal additive manufacturing (HMAM), 4–7
hydraulic transport systems and butterfly valve damage, 85–89

I

implementation of Industry 4.0 in multinational corporations (MNCs) 69
industrial applications, 10, 44
industrial automation, *182*
industrial development strategies and industry 4.0, 60–70
Industrial Internet of Things (IIoT), 64
industrial revolution, 69
Industry 4.0, 59, 69, 70, 76, 86
Industry 5.0 aspect, *130*
Industry 5.0 revolutions, *130*
injection molding and Bakelite product manufacturing, 179–183
intelligent welding systems in smart manufacturing, 60–65
Internet of Things (IoT), 59, 60, 64, 65, 69–73
Internet of Things (IoT) applications in manufacturing, 60–66

L

lattice Boltzmann method for ferrofluid heat transfer, 142–144
lean manufacturing and value stream mapping (VSM) 169–176
lean manufacturing system (LMS) 169
Levenberg–Marquardt algorithm, 149
life cycle and environmental impact of solar panels 132–134
literature review on plastic waste in concrete and bitumen, 154–160
literature review, 127–128, 187
logistics efficiency, 84
lubrication management in multistage forming, 62–63

M

manufacturing equipment, 7–11
manufacturing execution systems (MES), 60–61
manufacturing process simulation, 7
martensitic stainless steel *29*
material deposition, 13
material properties and microstructure of WAAMed parts, 21–37
medical industry, 4
mesh and simulation modeling in butterfly valve study, 85–89
metal powders, 7
microhardness and tensile strength in WAAM, 28
multiphase flow and CFD-DEM modeling, 85–89
multistage deep drawing of titanium in smart manufacturing, 74

N

nanoparticles and magnetic nanofluids, 142–146
novelty and innovation in product design, 179–183
numerical simulation of heat transfer and flow, 142–146
NX Siemens software 14, 15

O

open innovation in new product development, 180
operational demands in Bakelite product manufacturing, 185–188
orthogonal array Taguchi method, 40–48

P

packaging and handling in lubricant company study 172
passive cooling systems using ferrofluids 142–151
patentometric study on sugarcane juice extraction technologies, 113–124
plastic waste use in concrete and asphalt, 154–167
polymer matrix nanofillers (graphene), 130
post-processing, 7
pressure and velocity in butterfly valve flow, 85, 86, 87, 88, 89, 90, 91
process monitoring and control in smart manufacturing, 60–65
process parameter optimization, 40–55
process planning software in hybrid manufacturing, 14
product development and market success, 179–182
product homogeneity and ultrasonic pulse velocity (UPV), 101–107
production capacity in lubricant manufacturing, 176
productivity enhancement in smart manufacturing, 59
professionals shortage, 10, 11, 15, 16

Q

quality control 60, 70, 72, 74
quality control and monitoring, 8

R

raw material inventory and production process, 170
real-time data collection, 61, 62, 70, 71, 73
recycled HDPE tiles with graphene 126–139
recycling and sustainability in construction materials, 99–105
regression analysis in parameter optimization, 48–50
research and development, 16, 17
resin and plastic materials in construction, 154–157
return on investment (ROI), 77–80
road construction using plastic waste, 154–158
robotic welding systems, 60–65
ROMI D800 hybrid machining center, *13*
ROMI hybrid machine, 13, 14

S

sensor 22, 73
siemens NX, 14, 15
simulation studies (COMSOL, ANSYS Fluent), 142–151
smart manufacturing systems (SMS), Industry 4.0, 60–70
social innovation and social technology, 126, 127, 129, 131–134
software development, 10–15
solar energy and photovoltaic panels controversies, 132–134
solidification microstructures in 431, 29
specific software, 10
spray engineering and patent applicants, 121–123
strength properties of concrete, 154
surface integrity, 7
surface resistivity and water absorption in SCBA concrete, 107, 108
surface roughness optimization, 40–48
sustainability, 7, 18
sustainable construction materials, 99–105

T

takt time and lean production 169–176
teaching learning-based optimization (TLBO), 40–48
technological advancements, 3, 75, 117
technological improvement of 4.0 to Industry 5.0, *137*
technological prospecting and patentometry, 114, 117
technoscience as social technology, 126–132
thermal properties in heat sink design, 142
thermomagnetic convection in ferrofluid cooling, 142–152
training professionals, 4, 10, 11, 16
turbulence models in CFD analysis (k–ω SST) 89
types of sensors in IoT, 71

V

value stream map (VSM), 168, *170*
velocity magnitude comparison maximum value (m/s) data, 91

W

waste management in plastic and construction materials 154–159
waste plastic use in concrete and road construction, 154–167
water absorption in SCBA concrete, 108
welding and joining in smart manufacturing, 60–67
wire and powder materials, 6, 8
wire arc additive manufacturing (WAAM), 21–37
wire electrical discharge machining (WEDM), 40
workability and compressive strength of SCBA concrete, 103–105
workflow optimization using value stream mapping (VSM), 169–176
workforce skills, 10, 16, 18, 50, 63, 81

Made in the USA
Las Vegas, NV
19 September 2026

ada7a214-b82d-416d-801f-23f33c9772c3R01